HEINEMANN CHEMISTRY 1

SKILLS AND ASSESSMENT

Elissa Huddart

VCE Units 1 and 2

Written for the VCE Chemistry Study Design 2023–2027

Pearson Australia
(a division of Pearson Australia Group Pty Ltd)
459–471 Church Street
Level 1, Building B
Richmond, Victoria 3121
www.pearson.com.au

First published 2023 by Pearson Australia
2026 2025 2024 2023
10 9 8 7 6 5 4 3 2 1

VCE Heinemann Project Leads: Bryonie Scott, Misal Belvedere, Malcolm Parsons
Senior Development Editor: Fiona Cooke
Development Editor: Kate Rears
Schools Programme Manager: Michelle Thomas
Senior Production Editor: Casey McGrath
Editor: Catherine Greenwood
Series Designer: Anne Donald
Rights & Permissions Editor: Amirah Fatin Binte Mohamed Sapi'ee
Production Services Analyst: JitPin Chong
Design Analyst: Jennifer Johnston
Illustrator/s: DiacriTech and QBS Learning
Proofreader: Lorna Hendry
Printed in Australia by Pegasus Media and Logistics

ISBN 978 0 6557 0015 9

Pearson Australia Group Pty Ltd ABN 40 004 245 943

Indigenous Australians
We respectfully acknowledge the traditional custodians of the lands upon which the many schools throughout Australia are located. We acknowledge traditional Indigenous Knowledge systems are founded upon a logic that developed from Indigenous experience with the natural environment over thousands of years. This is in contrast to Western science, where the production of knowledge often takes place within specific disciplines. As a result, there are many cases where such disciplinary knowledge does not reflect Indigenous worldviews, and can be considered offensive to Indigenous peoples.

Some of the images used in *Heinemann Chemistry 1 Skills and Assessment* might have associations with deceased Indigenous Australians. Please be aware that these images might cause sadness or distress in Aboriginal or Torres Strait Islander communities.

Practical activities
All practical activities, including the illustrations, are provided as a guide only and the accuracy of such information cannot be guaranteed. Teachers must assess the appropriateness of an activity and take into account the experience of their students and facilities available. Additionally, all practical activities should be trialled before they are attempted with students and a risk assessment must be completed. All care should be taken and appropriate personal protective clothing and equipment should be worn when carrying out any practical activity. Although all practical activities have been written with safety in mind, Pearson Australia and the authors do not accept any responsibility for the information contained in or relating to the practical activities, and are not liable for loss and/or injury arising from or sustained as a result of conducting any of the practical activities described in this book.

Attributions
We thank the following for their contributions to our skills and assessment book:

The following abbreviations are used in this list: t = top, b = bottom, l = left, r = right, c = centre.

Cover: Science Photo Library: Thomas Deerinck, NCMIR

Science Photo Library: Andrew Lambert Photography, pp. 125tl, 125tr; Tompkinson, Geoff, pp. 1, 2, 52–4, 104–5; Turtle Rock Scientific/Science Source, pp. 108–9, 151, 198–9.

Shutterstock.com: Melnik, Vladimir, p. 118; Monticello, p. 22.

The Royal Society of Chemistry: Practical procedure © Nuffield Foundation and The Royal Society of Chemistry, reproduced by permission, source https://edu.rsc.org/experiments/making-plastic-from-potato-starch/1741.article, pp. 98-9.

vic.gov.au: © Copyright State Government of Victoria. Licensed under Creative Commons Attribution 4.0 International licence, link to licence: https://creativecommons.org/licenses/by/4.0/legalcode, p. 107.

Victorian Curriculum and Assessment Authority (VCAA): Selected extracts from the VCE (Chemistry) Study Design (2023–2027) are copyright. Victorian Curriculum and Assessment Authority (VCAA), reproduced by permission. VCE® is a registered trademark of the VCAA. The VCAA does not endorse this product and makes no warranties regarding the correctness or accuracy of its content. To the extent permitted by law, the VCAA excludes all liability for any loss or damage suffered or incurred as a result of accessing, using or relying on the content. Current VCE Study Designs and related content can be accessed directly at www.vcaa.vic.edu.au, pp. 1–2, 53–4, 105, 109, 151, 199.

CHEMISTRY TOOLKIT viii

Unit 1 How can the diversity of materials be explained?

AREA OF STUDY 1

How do the chemical structures of materials explain their properties and reactions?

KEY KNOWLEDGE 2

WORKSHEETS

WORKSHEET 1	Knowledge review—structure of the atom	15
WORKSHEET 2	Writing electronic configurations—shells and subshells	16
WORKSHEET 3	Patterns in properties in the periodic table	17
WORKSHEET 4	Representations of molecules	18
WORKSHEET 5	Electronegativity and polarity of molecules	19
WORKSHEET 6	The metallic bonding model	20
WORKSHEET 7	The ionic bonding model	21
WORKSHEET 8	Writing ionic formulas	22
WORKSHEET 9	Solubility tables and predicting precipitation reactions	23
WORKSHEET 10	Writing full and ionic chemical equations	24
WORKSHEET 11	Literacy review—comparing similar terms	25
WORKSHEET 12	Reflection—How do the chemical structures of materials explain their properties and reactions?	26

PRACTICAL ACTIVITIES

ACTIVITY 1	Using flame colours to identify elements	27
ACTIVITY 2	Making molecular models	30
ACTIVITY 3	Comparing the physical properties of three covalent lattices	33
ACTIVITY 4	Growing metal crystals	36
ACTIVITY 5	Reactivity of metals—student-designed practical activity	38
ACTIVITY 6	Precipitation reactions	40
ACTIVITY 7	Chromatography of a vegetable extract	45

EXAM-STYLE QUESTIONS 48

AREA OF STUDY 2

How are materials quantified and classified?

KEY KNOWLEDGE 54

WORKSHEETS

WORKSHEET 13	Knowledge review—comparing metallic, ionic and covalent bonding models	66
WORKSHEET 14	Exploring relative mass	67
WORKSHEET 15	Moles—the chemist's unit of measurement	68
WORKSHEET 16	Empirical and molecular formulas	69
WORKSHEET 17	Alkanes, alkenes and haloalkanes	70
WORKSHEET 18	Families of organic molecules—haloalkanes, alcohols and carboxylic acids	72
WORKSHEET 19	Polyethene—a case study of a polymer	73
WORKSHEET 20	Designing a polymer for a particular purpose	75
WORKSHEET 21	Literacy review—naming compounds	76
WORKSHEET 22	Reflection—How are materials quantified and classified?	77

PRACTICAL ACTIVITIES

ACTIVITY 8	Mole simulations and applications	78
ACTIVITY 9	Determining the molar mass of an element and a compound	81
ACTIVITY 10	Chemical composition of a compound	86
ACTIVITY 11	Investigating hydrocarbons	89
ACTIVITY 12	Modelling functional groups	92
ACTIVITY 13	Investigating properties of slime, an addition polymer	95
ACTIVITY 14	Making a bioplastic	98

EXAM-STYLE QUESTIONS 100

AREA OF STUDY 3

How can chemical principles be applied to create a more sustainable future?

RESEARCH INVESTIGATION	Investigating the issue of banning single-use plastics in Victoria	106

Unit 2 How do chemical reactions shape the natural world?

AREA OF STUDY 1

How do chemicals interact with water?

KEY KNOWLEDGE 110

WORKSHEETS

WORKSHEET 23 Knowledge review—identifying and naming types of substances, balancing chemical equations 117
WORKSHEET 24 Structure and properties of water 118
WORKSHEET 25 Calculations using specific heat capacity 119
WORKSHEET 26 Calculations using latent heat 120
WORKSHEET 27 Concentration and strength—picturing acids and bases 121
WORKSHEET 28 Predicting products of acid reactions 122
WORKSHEET 29 Calculating pH 123
WORKSHEET 30 Redox reactions and reactivity of metals 125
WORKSHEET 31 Literacy review—matching redox key terms 127
WORKSHEET 32 Reflection—How do chemicals interact with water? 128

PRACTICAL ACTIVITIES

ACTIVITY 15 Density of water and ice 129
ACTIVITY 16 Investigating acids 131
ACTIVITY 17 Reactions of HCl with metals and carbonates 135
ACTIVITY 18 Beetroot—a natural indicator 138
ACTIVITY 19 Reactivity series of metals 141
ACTIVITY 20 Comparing a simple primary cell and a direct reaction 144

EXAM-STYLE QUESTIONS 147

AREA OF STUDY 2

How are chemicals measured and analysed?

KEY KNOWLEDGE 152

WORKSHEETS

WORKSHEET 33 Knowledge review—elements, compounds and molar mass 161
WORKSHEET 34 Molarity—measuring moles in solution 162
WORKSHEET 35 Converting between concentration units 163
WORKSHEET 36 Removing impurities from water using precipitation reactions 164
WORKSHEET 37 Standard solutions 165
WORKSHEET 38 Acid–base titrations 166
WORKSHEET 39 Mass–volume stoichiometry for gases 167
WORKSHEET 40 Solving complex calculations—using more than one formula 168
WORKSHEET 41 Analysis with light—colorimetry and UV–visible spectroscopy 169
WORKSHEET 42 Literacy review—key terms and formulas 171
WORKSHEET 43 Reflection—How are chemicals measured and analysed? 172

PRACTICAL ACTIVITIES

ACTIVITY 21 Determination of solubility of a salt in water 173
ACTIVITY 22 Preparation of a standard solution 176
ACTIVITY 23 Determination of HCl content in brick cleaner 178
ACTIVITY 24 Investigating the volume–pressure relationship in gases 181
ACTIVITY 25 Determining the molar volume of hydrogen 184
ACTIVITY 26 Gravimetric determination of sulfur as sulfate in fertiliser 187
ACTIVITY 27 Colorimetric determination of phosphorus content of lawn fertiliser 190

EXAM-STYLE QUESTIONS 194

AREA OF STUDY 3

How do quantitative scientific investigations develop our understanding of chemical reactions?

SCIENTIFIC INVESTIGATION Investigating the concentration of ethanoic acid in different types of vinegar 200

How to use this book

The *Heinemann Chemistry 1 Skills and Assessment* book provides the opportunity to practise, apply and extend your learning through a range of supportive and challenging activities. These activities reinforce key concepts and skills, and enable a flexible approach to learning. There are also regular opportunities for reflection and self-evaluation in the final worksheet in each area of study.

This resource has been written to the VCE Chemistry Study Design 2023–2027 and is divided into six areas of study—three in Unit 1 and three in Unit 2. Areas of Study 1 and 2 in each unit consist of four main sections:

- key knowledge
- worksheets
- practical activities
- exam-style questions.

CHEMISTRY TOOLKIT

The Chemistry toolkit supports development of the skills and techniques needed to undertake practical and secondary-sourced investigations, and covers examination techniques and study skills. It also includes checklists, models, exemplars and scaffolded steps. The toolkit can serve as a reference tool and be consulted as needed.

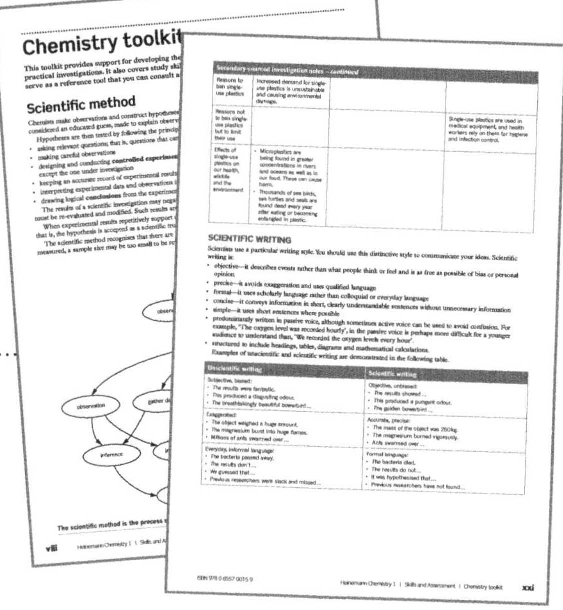

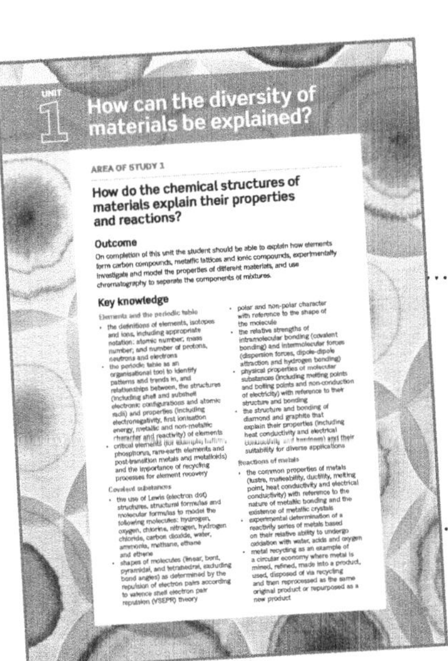

AREA OF STUDY OPENER

Heinemann Chemistry 1 Skills and Assessment is structured to follow the study design units and areas of study. The area of study opening page lists the study design key knowledge for easy reference to the activities that follow.

KEY KNOWLEDGE

Each area of study begins with a key knowledge section. This consists of a set of summary notes that cover the key knowledge for that area of study. Key terms are in bold and are included in the glossary of the student book. The section also serves as a ready reference for completing the worksheets and practical activities.

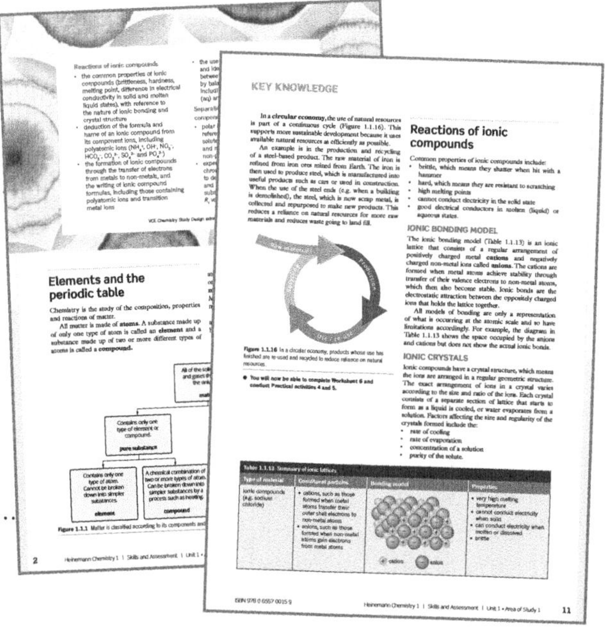

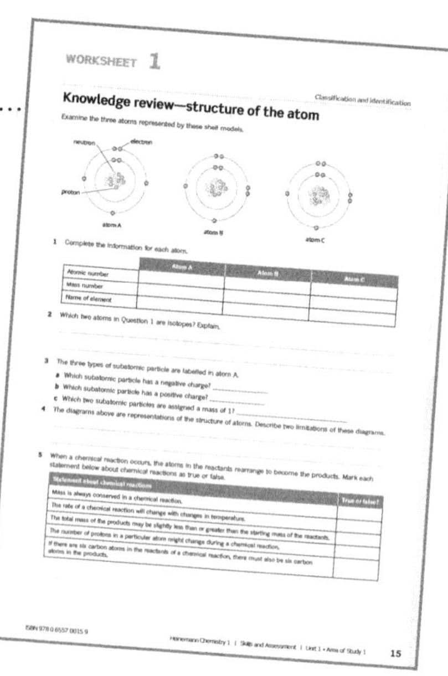

WORKSHEET 1

Classification and identification

Knowledge review—structure of the atom

Examine the three atoms represented by these shell models.

WORKSHEETS

The worksheets feature questions that allow you to practise and apply your knowledge and skills. Each area of study includes a 'Knowledge review' worksheet to activate prior knowledge, a 'Literacy review' worksheet that provides opportunities for vocabulary and literacy support, and a 'Reflection' worksheet, which you can use for self-assessment. Other worksheets provide opportunities to revise, consolidate and further your understanding. All worksheets function as formative assessment and are clearly aligned with the study design. A range of questions building from foundation to challenging is included in each worksheet.

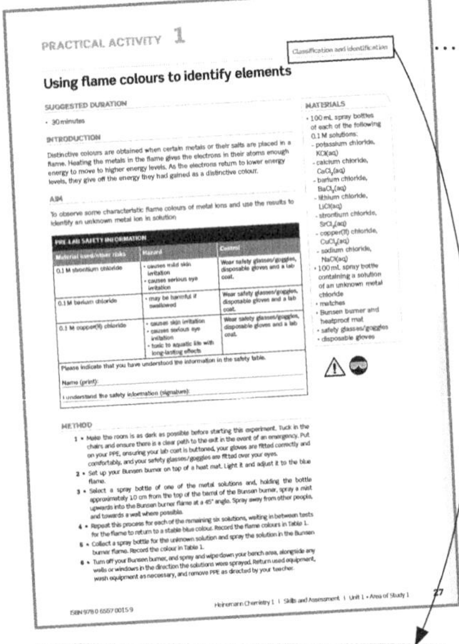

PRACTICAL ACTIVITY 1

Classification and identification

Using flame colours to identify elements

SUGGESTED DURATION

INTRODUCTION

AIM

MATERIALS

PRE-LAB SAFETY INFORMATION

METHOD

Classification and identification

PRACTICAL ACTIVITIES

Practical activities offer you the chance to complete practical work related to the various themes covered in the study design. You have the opportunity to design and conduct scientific investigations, generate, evaluate and analyse data, appropriately record results, and prepare evidence-based conclusions. Where relevant, you will also need to conduct risk assessments to identify any potential hazards.

Each practical activity includes a suggested duration. Together with the Area of Study 3 scientific investigations, the practical activities meet the 34 hours of practical work mandated for Units 1 and 2 in the study design.

METHODOLOGIES

Each worksheet and practical activity is mapped to one or more of the scientific investigation methodologies outlined in the study design. Completing these activities gives you experience in applying the methodologies in a wide variety of contexts and prepares you for designing and conducting your own scientific investigation in Unit 2 Area of Study 3.

EXAM-STYLE QUESTIONS

Each area of study finishes with a selection of exam-style questions. These give you the opportunity to gain valuable experience in applying your knowledge and understanding to exam-style questions.

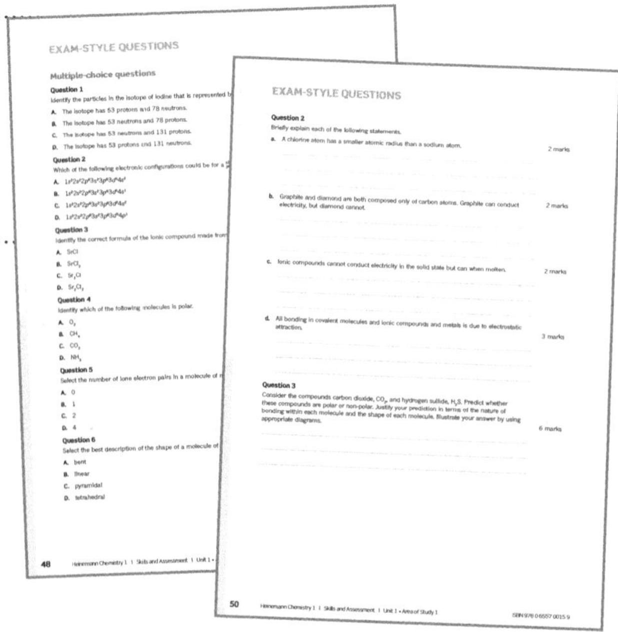

EXAM-STYLE QUESTIONS

Multiple-choice questions

EXAM-STYLE QUESTIONS

TEACHER SUPPORT

Comprehensive answers and fully worked solutions for all worksheets, practical activities and exam-style questions are provided via the *Heinemann Chemistry 1 eBook + Assessment* or Pearson Places. In-depth support for Unit 1 Area of Study 3 and Unit 2 Area of Study 3 in the form of samples, templates and teacher notes is also included, along with an interactive SPARKlab for every practical activity.

ISBN 978 0 6557 0015 9

Chemistry toolkit

This toolkit provides support for developing the skills required to undertake research and practical investigations. It also covers study skills and examination preparation. The toolkit can serve as a reference tool that you can consult as needed throughout the year.

Scientific method

Chemists make observations and construct hypotheses to account for their observations. A **hypothesis** can be considered an educated guess, made to explain observations.

Hypotheses are then tested by following the principles of the **scientific method**. These include:

- asking relevant questions; that is, questions that can be tested
- making careful observations
- designing and conducting **controlled experiments**; in controlled experiments all **variables** are kept constant, except the one under investigation
- keeping an accurate record of experimental results
- interpreting experimental data and observations in a logical manner
- drawing logical **conclusions** from the experimental results.

The results of a scientific investigation may negate or refute the hypothesis being tested. In this case, the hypothesis must be re-evaluated and modified. Such results are useful in redirecting scientific investigation.

When experimental results repetitively support a hypothesis, the hypothesis may become a **theory** or **principle**; that is, the hypothesis is accepted as a scientific truth.

The scientific method recognises that there are limitations in investigations. For example, some factors cannot be measured, a sample size may be too small to be representative, or unknown factors may influence investigations.

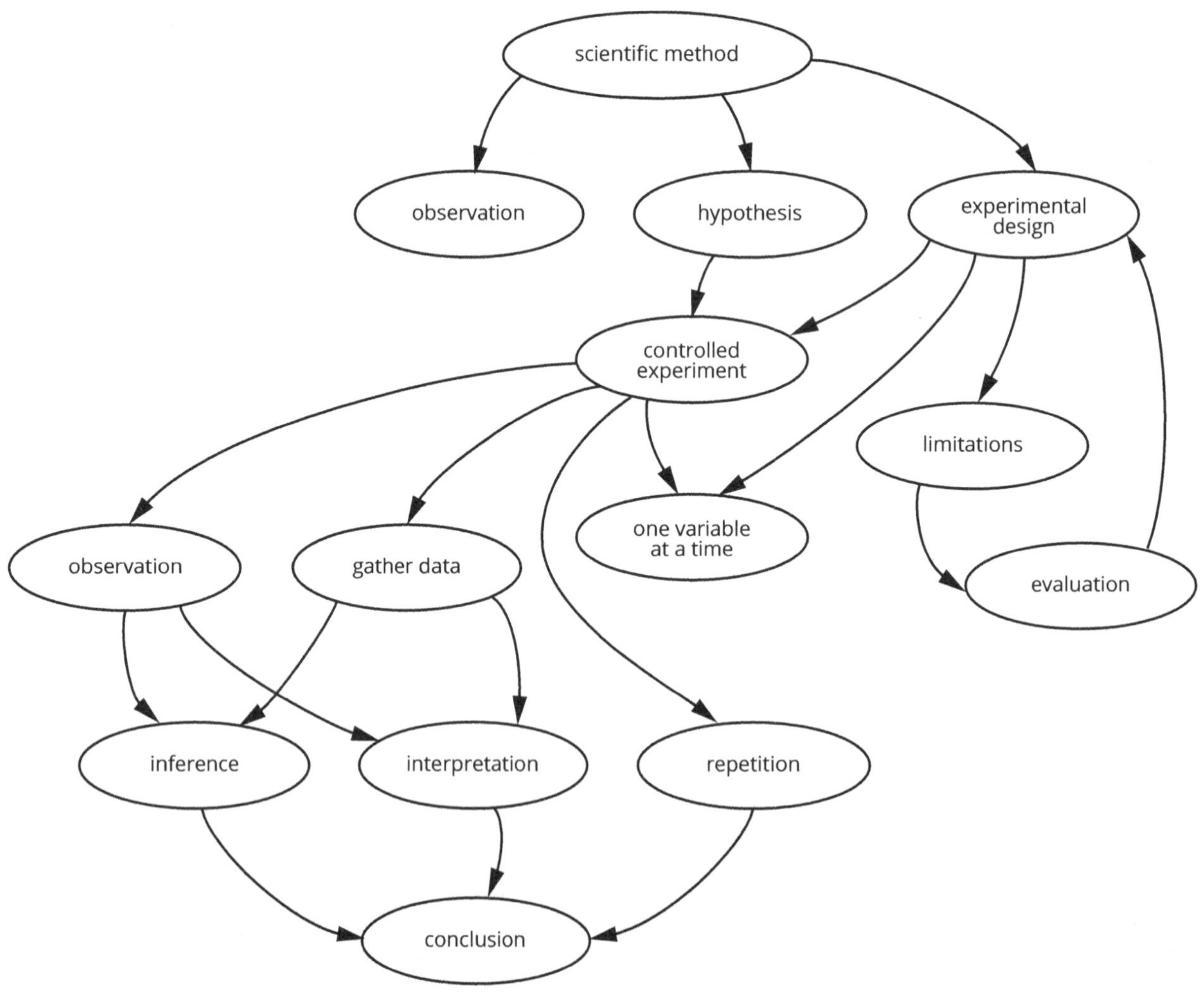

The scientific method is the process scientists use to conduct experiments and test hypotheses.

ISBN 978 0 6557 0015 9

SCIENTIFIC INVESTIGATION

A scientific investigation can be conducted in many ways. Examples of scientific investigation approaches (also known as methodologies) are controlled experiments, literature reviews and modelling. The scientific investigation methodology and the methods (also known as procedures) selected will depend on the aim of the investigation and the research question.

Practical work (or practical activities) involves direct experiences or hands-on activities. Scientific investigation methodology involves all elements of planning—it considers the focus of the investigation and the rationale for approaching investigations in a particular way; for example, through controlled experiments, fieldwork or modelling. See the checklist below and on page x for further information about methodology versus method in scientific investigations.

An investigation that you conduct yourself is known as a primary investigation, and the data and information you collect is called primary data or a primary source.

An investigation in which you analyse data collected by others is known as a secondary-sourced investigation (see page xix).

When planning and conducting a scientific investigation, you must maintain a logbook to record information related to your investigation, such as materials and methods, raw data, data analysis and sources of information.

The findings of a scientific investigation may be presented in a variety of formats, such as a scientific report, an article or a scientific poster.

Conducting a scientific investigation

Scientific investigations follow a precise scientific method. The checklist below and on page x provides a summary of the elements that are common to many scientific investigation methodologies and scientific reports. Refer to the checklist and record important information as you conduct your scientific investigation.

Element	Explanation	Tick ✔
date	• the date(s) the scientific investigation was conducted, and the report was written	
title/heading	• the question under investigation	
aim	• a statement that describes what the scientist wants to investigate, demonstrate or find out • can be written as 'to investigate the effect of *x* on *y*'	
hypothesis	• a tentative explanation for an observation that is based on evidence and prior knowledge • must be testable • written as a short statement • should include variables to be tested and a statement of the measurable, predicted outcome • can be written as 'If *x* happens, then *y* will happen'. The 'if' part relates to the independent variable. The 'then' part relates to the dependent variable.	
variables	• experiments measure relationships between variables • the independent variable is the variable that is changed • the dependent variable is the variable responding to the change in the independent variable • the controlled variables are kept constant throughout the experiment	
risk assessment/ safety	• an analysis of the potential risks of conducting the investigation • objective is to maximise safety by managing and minimising risk (the risk assessment form on page xi guides you through the analysis of risk)	
materials	• a list of all equipment, chemicals and materials used • includes quantity, volume, concentration and mass of materials used	
methodology	• a description of the overall approach undertaken in the scientific investigation and the reasons for taking this approach	
method	• a detailed, numbered, step-by-step list of exactly what is to be done • includes details about the materials used at each step (e.g. volume and concentration of a chemical) • may include scientific drawings or labelled diagrams • includes enough detail to allow the experiment to be repeated exactly • considers steps that can be taken to reduce errors	
results	• a record of all observations and measurements taken during the investigation • observations may be recorded as text, diagrams, photos or videos • data is commonly presented in tables and graphs and can include calculations • data should be checked for errors	

continues over

Element	Explanation	Tick ✔
discussion	• interprets data, and links results to chemical concepts • includes analysis of the data, looking for trends and relationships • identifies problems such as bias, inaccuracy, unreliable data and errors • evaluates the method used, including a discussion of limitations • comments on whether the results do or do not support the hypothesis • comments on the implications of the investigation and further investigations that may be undertaken	
conclusion	• a short statement based on evidence that summarises the findings of the investigation • provides a response to the research question or hypothesis • identifies the extent to which the investigation addressed the research question or hypothesis • does not introduce any new information	

RISK ASSESSMENT

Five levels of safety should be considered before carrying out a scientific investigation. The inverted pyramid ranks these levels in order of importance. The school and your teacher are responsible for reducing most of these risks. As a student scientist, you can take measures to reduce the risks shown at the bottom of the hierarchy.

To determine the risks involved in an investigation, you should complete a risk assessment beforehand. Complete the risk assessment form on page xi to identify possible risks for which you can take responsibility, and to think of ways you can reduce risks to create a safe environment.

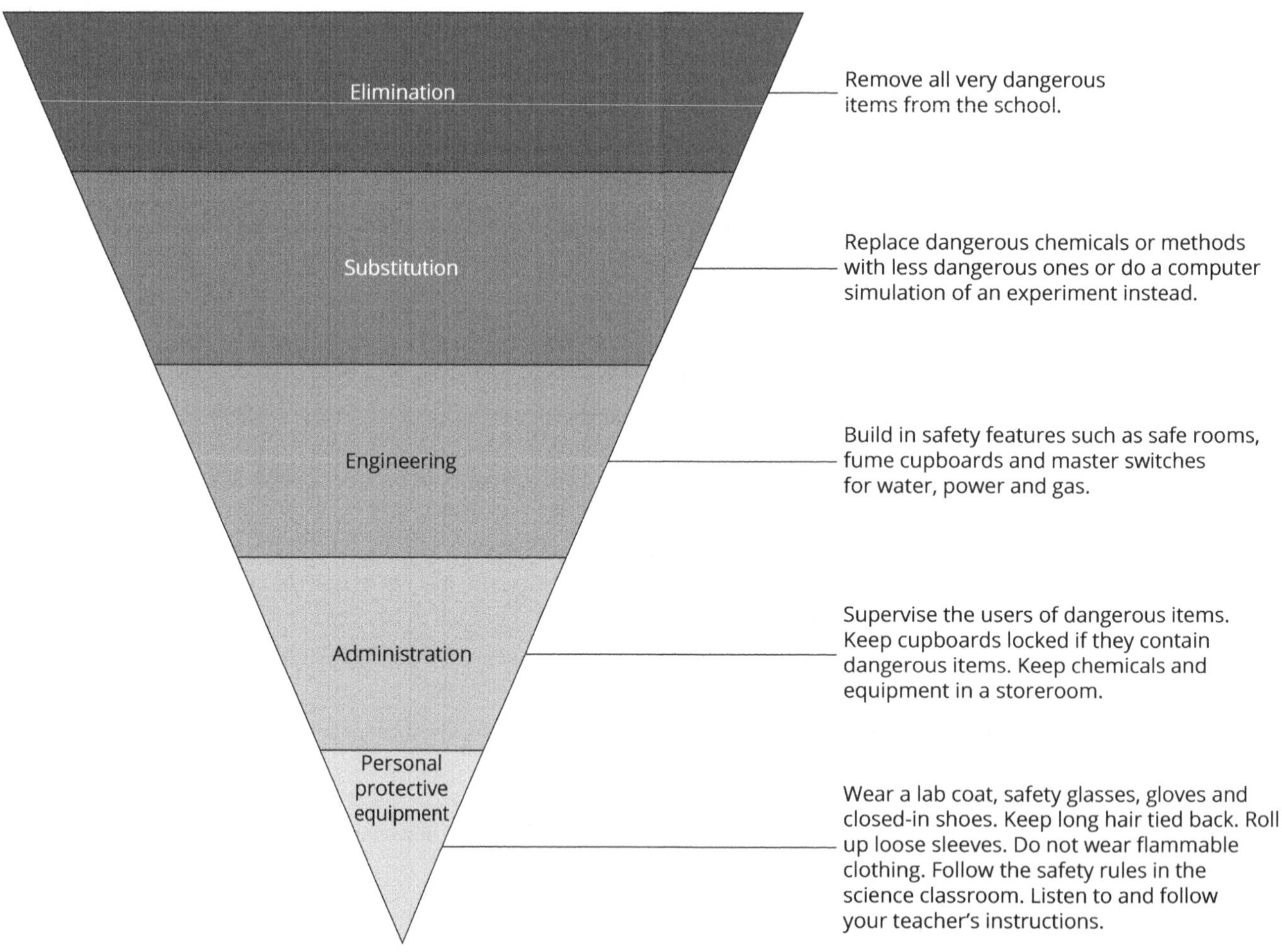

 ISBN 978 0 6557 0015 9

Risk assessment form		
What activity are you doing?		
Title or description of the investigation:		
List ...	**Identify any risks**	**State how you will ...**
equipment you will be using:		use each piece:
chemicals you will be using:		carefully use each chemical: carefully dispose of the chemicals:
ethical issues you need to consider:		ethically use animals in the laboratory: ethically use human participants in the investigation:
outdoor or fieldwork activities:		reduce these risks:
any other possible risks:		reduce these risks:

EXAMPLES OF SCIENTIFIC REPORTS

It can be difficult to gauge whether you have attained a high standard in your completed scientific report. Looking at sample scientific reports can help you identify what is required. Two sample scientific reports are provided here as a reference: one is prepared to a high standard and the second has room for improvement. They include annotations to draw your attention to key points to note on each scientific report. These points are also reflected in the checklist, so you can use this as a tool to evaluate whether all requirements of the scientific investigation are complete.

High standard practical report

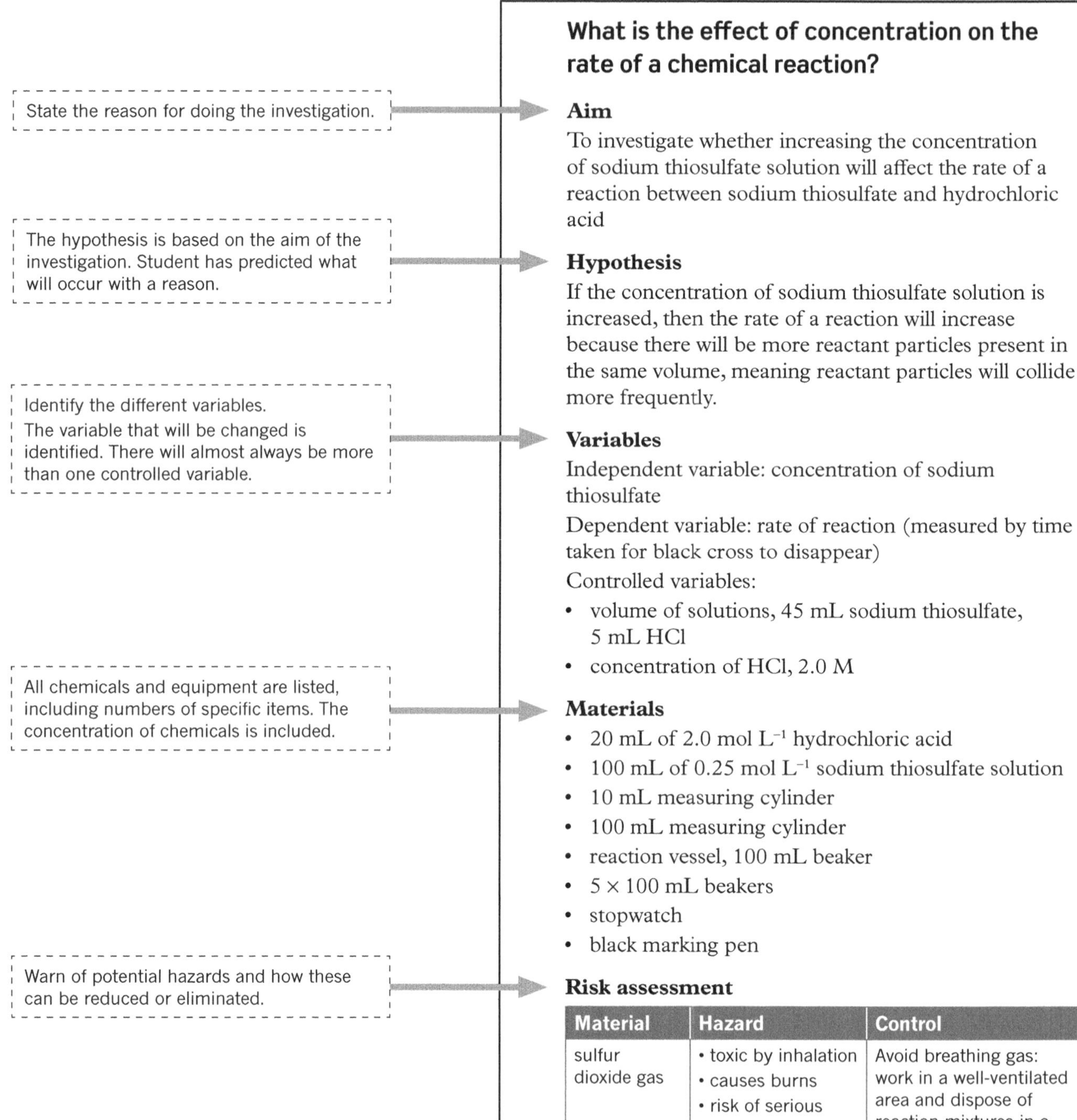

What is the effect of concentration on the rate of a chemical reaction?

Aim

To investigate whether increasing the concentration of sodium thiosulfate solution will affect the rate of a reaction between sodium thiosulfate and hydrochloric acid

Hypothesis

If the concentration of sodium thiosulfate solution is increased, then the rate of a reaction will increase because there will be more reactant particles present in the same volume, meaning reactant particles will collide more frequently.

Variables

Independent variable: concentration of sodium thiosulfate

Dependent variable: rate of reaction (measured by time taken for black cross to disappear)

Controlled variables:

- volume of solutions, 45 mL sodium thiosulfate, 5 mL HCl
- concentration of HCl, 2.0 M

Materials

- 20 mL of 2.0 mol L^{-1} hydrochloric acid
- 100 mL of 0.25 mol L^{-1} sodium thiosulfate solution
- 10 mL measuring cylinder
- 100 mL measuring cylinder
- reaction vessel, 100 mL beaker
- 5 × 100 mL beakers
- stopwatch
- black marking pen

Risk assessment

Material	Hazard	Control
sulfur dioxide gas	• toxic by inhalation • causes burns • risk of serious damage to eyes	Avoid breathing gas: work in a well-ventilated area and dispose of reaction mixtures in a sink in a fume cupboard. Wear safety glasses.
sulfur powder	highly flammable	Keep away from sources of ignition. Do not breathe dust. Avoid contact with eyes.
2.0 mol L^{-1} hydrochloric acid	splashes to eyes	Wear safety glasses.

ISBN 978 0 6557 0015 9

Methodology

A controlled experiment was conducted to determine the effect of concentration on the rate of a reaction between sodium thiosulfate and hydrochloric acid.

Describe the overall approach undertaken in the scientific investigation and the reasons for taking this approach.

Method

Provide clear instructions about each step of the experiment. These should be written in the third person as recipe-style, numbered, easy-to-follow detailed instructions.

1 Mark a cross on a sheet of paper with a black marking pen.
2 Place a 100 mL beaker on top of the cross. Pour 10 mL of 0.25 mol L^{-1} sodium thiosulfate solution and 35 mL of water into the beaker. Add 5 mL of 2 mol L^{-1} hydrochloric acid and commence timing. Measure and record in Table 1 the time taken for the cross to disappear when viewed from above the beaker.
3 Place a 100 mL beaker on top of the cross. Pour 15 mL of 0.25 mol L^{-1} sodium thiosulfate solution and 30 mL of water into the beaker. Add 5 mL of 2 mol L^{-1} hydrochloric acid and commence timing. Measure and record the time taken for the cross to disappear when viewed from above the beaker.
4 Place another 100 mL beaker on top of the cross. Pour 25 mL of the sodium thiosulfate solution and 20 mL of water into the beaker. Add 5 mL of hydrochloric acid and commence timing. Measure and record the time taken for the cross to disappear.
5 Place a 100 mL beaker on top of the cross. Pour 35 mL of 0.25 mol L^{-1} sodium thiosulfate solution and 10 mL of water into the beaker. Add 5 mL of 2 mol L^{-1} hydrochloric acid and commence timing. Measure and record the time taken for the cross to disappear when viewed from above the beaker.
6 Place another 100 mL beaker on top of the cross. Pour 40 mL of sodium thiosulfate solution and 5 mL of water into the beaker. Add 5 mL of hydrochloric acid and commence timing. Measure and record the time taken for the cross to disappear.
7 Calculate a reaction rate according to reaction rate $= \frac{1}{\text{time}}$ taken for cross to disappear and add to Table 1.

Raw and processed data are included.

Tables have a number and title.

Columns have a title and unit.

This is a record of all observations and measurements taken during the investigation.

Graphs, diagrams and photographs are also useful approaches for recording data.

The dependent variable is plotted on the *y*-axis.

Axes are labelled with a title and unit.

Graphs are numbered.

Results

Table 1 Results for effect of concentration of reactants

Solution	Concentration of sodium thiosulfate in reactant solution (mol L^{-1})	Concentration of HCl (mol L^{-1})	Time for the cross to disappear (s)	Reaction rate (s^{-1})
10 mL $Na_2S_2O_3$ solution, 35 mL water and 5 mL HCl	0.05	2.0	80	0.013
15 mL $Na_2S_2O_3$ solution, 30 mL water and 5 mL HCl	0.075	2.0	51	0.020
25 mL $Na_2S_2O_3$ solution, 20 mL water and 5 mL HCl	0.125	2.0	28	0.036
35 mL $Na_2S_2O_3$ solution, 10 mL water and 5 mL HCl	0.175	2.0	18	0.056
40 mL $Na_2S_2O_3$ solution, 5 mL water and 5 mL HCl	0.20	2.0	5	0.091

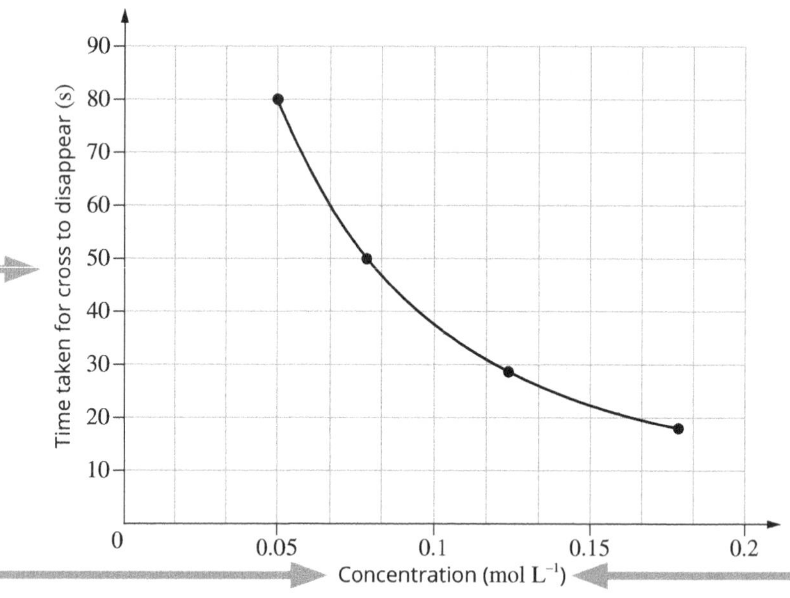

Figure 1 Time taken for cross to disappear versus concentration of sodium thiosulfate solution

Graphs have a title that is descriptive.

The independent variable is plotted on the *x*-axis.

ISBN 978 0 6557 0015 9

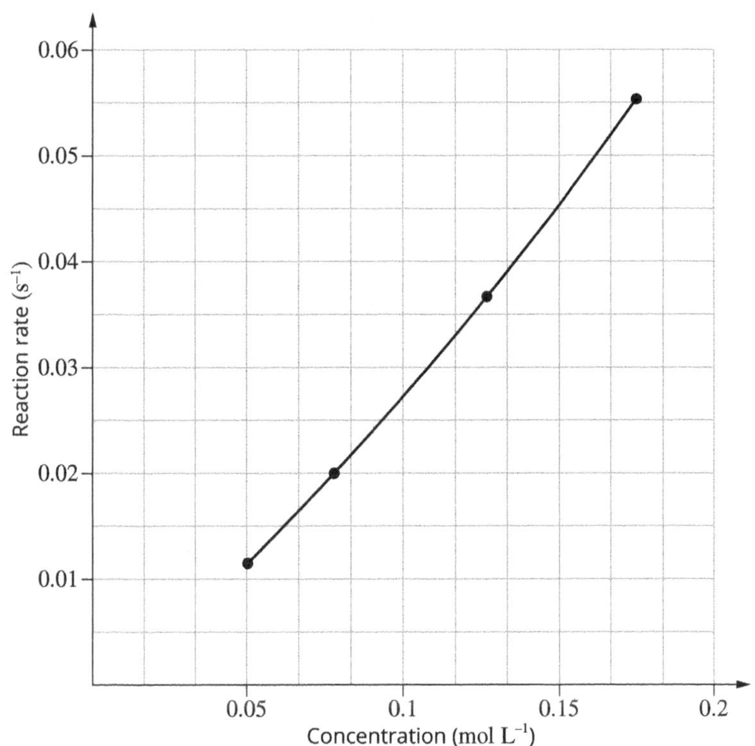

Figure 2 Reaction rate versus concentration of sodium thiosulfate solution

Discussion

The discussion focuses on the interpretation of the experimental data. What do the results show? Are there any unexpected results? How can we account for this? The discussion also includes an evaluation of the procedure.

As can be seen in Figure 2, the rate of reaction with a higher concentration of sodium thiosulfate was higher than with lower concentrations. There is evidence of a direct relationship between concentration and rate.

This observation is probably because the increased concentration of reactant particles led to a greater frequency of collisions. This in turn means successful collisions occurred more frequently and the reaction occurred faster.

One piece of data for the $Na_2S_2O_3$ concentration of 0.20 mol L^{-1} was considered to be an outlier. This error could be explained by the fact that the timer was not started until slightly after the reactants were combined. The reliability of the results would have been improved if each test was repeated three times in order to obtain an average time.

The experiment could also be improved by increasing the range of concentration of sodium thiosulfate to see if the curve is a straight line. An alternative experiment that varied the concentration of HCl would also support whether it is the concentration of either reactants that affects the reaction rate, or whether it is the concentration of sodium thiosulfate.

Conclusion

The conclusion relates back to the purpose and states whether the hypothesis was supported or not supported. Examples of supporting experimental evidence are included.

Avoid terms such as 'always' or 'never' in scientific writing. Refer to what the data shows—all claims must be supported by the evidence.

The data showed that the reaction mixtures with a higher concentration reacted faster than those with lower concentration. For example, 0.05 mol L^{-1} $Na_2S_2O_3$ had a reaction rate of 0.013 s^{-1} and 0.175 mol L^{-1} $Na_2S_2O_3$ had a higher rate of 0.056 s^{-1}. From this, it can be concluded that increasing concentration increases the rate of a chemical reaction; thus, the hypothesis has been supported.

Low standard practical report

The title is too general.

The aim is not specific enough and is not followed by a hypothesis.

More variables will need to be controlled and should be listed here.

Quantities and concentrations should be listed.

There should be details of how the risks will be controlled.

The selected methodology is not outlined.
The method must be a set of detailed step-by-step instructions that can be easily repeated by someone else.

Only testing two different concentrations of sodium thiosulfate is not enough to draw conclusions.

An investigation into reaction rates

Aim

To determine whether increasing the concentration will increase the rate of a reaction

Variable

Independent variable: concentration of sodium thiosulfate

Dependent variable: time taken for black cross to disappear

Controlled variables: volume of solutions

Materials

- hydrochloric acid
- sodium thiosulfate solution
- measuring cylinders
- 100 mL beakers
- thermometer
- stopwatch
- black marking pen

Risk assessment

Material	Hazard
sulfur dioxide gas	toxic by inhalation; causes burns; risk of serious damage to eyes
sulfur powder	highly flammable
2.0 mol L^{-1} hydrochloric acid	Beware of splashes to eyes.

Method

1 A cross was marked on a piece of paper. A beaker was placed on the cross.

2 10 mL of the sodium thiosulfate solution and 35 mL of water were placed into the beaker. 5 mL of 2 mol L^{-1} hydrochloric acid was added. Record the time taken for the cross to disappear.

ISBN 978 0 6557 0015 9

3 Repeat with 40 mL of sodium thiosulfate solution and 5 mL of water into the beaker. Add 5 mL of hydrochloric acid. The time taken for the cross to disappear was measured.

The tense changes in the writing.

Results

All tables should have a heading.

Solution	Concentration of sodium thiosulfate in reactant solution (mol L^{-1})	Concentration of HCl (mol L^{-1})	Time for the cross to disappear (s)
10 mL $Na_2S_2O_3$ solution, 35 mL water and 5 mL HCl	0.05	2.0	80
40 mL $Na_2S_2O_3$ solution,5 mL water and 5 mL HCl	0.20	2.0	5

Time taken for the cross to disappear is not the same as reaction rate. A further calculation would be better.

Discussion

The rate of reaction was faster for the higher concentration of sodium thiosulfate.

This observation was due to the increased concentration of reactant particles leading to a greater frequency of collisions.

This discussion is too brief—it doesn't provide detailed explanation to account for the results or reference to experimental data. Errors, improvements and an evaluation of the method should also be included.

Conclusion

The data showed that the reaction mixtures with a higher concentration reacted faster.

This conclusion is too vague. It doesn't respond specifically to the aim of the investigation, nor state whether the hypothesis was supported or not supported. It doesn't use the experimental results to support the conclusion.

STUDENT-DESIGNED INVESTIGATION

You will be required to design and conduct a scientific investigation based on the concepts you have learned in Unit 2, Areas of Study 1 and/or 2. This assessment task gives you the opportunity to apply key science skills and to pursue an area of interest to you, based on the key knowledge addressed during the course. You will be required to develop a question that drives your investigation, state an aim and hypothesis, select appropriate methodology and methods, and generate and collect primary quantitative data, recording important information in your logbook. You will then present your investigation in a format of your choice, or as guided by your teacher; for example, a practical report, a scientific poster, a multimedia presentation or a written article. It will be helpful to refer to pages xii–xiii to review what to include in a formal scientific report.

It will be important to carefully select the appropriate methodology and methods for your investigation. You will need to be clear about the difference between the two.

- Methodology describes the overall approach undertaken in a scientific investigation; it considers the investigation more broadly, and includes the reasons for taking the chosen approach; for example, it will identify and describe strategies, such as the methods used to obtain data and the reasons why this is important to achieve the aim of your scientific investigation and address the question under investigation.
- The method is the specific procedure or steps taken to collect data during a scientific investigation.

DETERMINING YOUR RESEARCH QUESTION

As you develop a question for scientific investigation, be aware of the depth of thinking it will require. The table on page xviii provides support in writing questions at different levels of thinking or complexity. It also provides some key words to help you target these thinking levels and gives some examples of questions.

When developing your question for investigation, be conscious of the level you are aiming for. Questions for scientific investigation should generally be at the analysis level.

Level of question complexity	Type of thinking	Words that might be used		Examples of questions and commands
Simple	**Retrieval:** • remembering, producing information on demand	• who • where • list • show • describe • select • complete • define	• what • when • label • demonstrate • name • state • recognise • identify	**1** Define 'molecule'. **2** What is the function of intermolecular forces? **3** List the metallic elements in order of least to most reactive. **4** What are the products of the reaction between acids and metal carbonates?
	Comprehension: • the ability to understand information	• who • where • explain • represent • show how	• what • when • summarise • draw • describe	**1** Represent the Bohr model of the atom with a labelled diagram. **2** Explain how the atomic radius of an oxygen atom is smaller than the atomic radius of a lithium atom. **3** Explain why the mole concept is useful to chemists.
	Application: • using knowledge in new situations, including: - testing a hypothesis - solving a problem - experimenting and using data - decision-making	• why • investigate • find out about • test • solve • develop • decide • construct	• how • research • experiment • predict • adapt • judge	**1** Many scientists believe that limiting the availability of single-use plastic bags is necessary to control environmental damage. Construct an argument for or against this statement. **2** Investigate the behaviour of a silicon diode in a simple circuit. **3** How would different temperatures of water affect the rate at which a submerged iron rod rusts?
Complex (requires more thinking)	**Analysis:** • scrutinising and breaking something into its smaller parts, including: - comparing - classifying - identifying errors - concluding - predicting - judging	• why • categorise • contrast • sort • organise • generalise • evaluate • edit • assess • judge	• how • compare • distinguish • discriminate between • deduce • critique • diagnose • identify errors • identify misunderstandings	**1** Categorise the following compounds according to their type of bonding. **2** Compare the properties and behaviour of gases and liquids. **3** Compare specific heat capacity and latent heat.

ISBN 978 0 6557 0015 9

SECONDARY-SOURCED INVESTIGATION

Some scientific investigations require the collation and analysis of data that others have collected. Data or information that was collected by someone else is known as secondary data or a secondary source. An investigation that uses secondary data is known as a secondary-sourced investigation. An example of a scientific investigation methodology that involves collating and analysing secondary sources is a literature review. This section guides you in conducting a secondary-sourced investigation.

Activities such as investigating the sustainability of a material are likely to rely on secondary sources of information. Such investigations require you to think carefully about the topic; find, collect and organise information; analyse and synthesise findings; and present your ideas.

The process for undertaking a secondary-sourced investigation is summarised in the following flow chart. A secondary-sourced investigation is not necessarily a straightforward linear process, as shown in the flow chart. You can move back and forth between steps as needed.

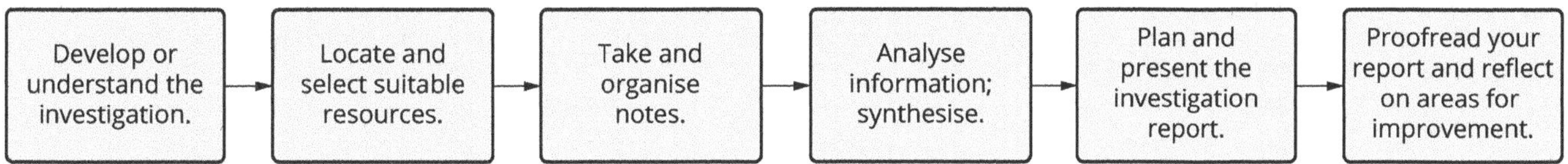

SOURCING INFORMATION

The resources you refer to in investigations may be primary and/or secondary.

Primary sources of information are created by a person directly involved in an investigation or a study. Examples of primary sources are results from experiments (such as raw data or photographs), reports of scientific investigations, specimens or artefacts collected, and peer-reviewed scientific articles reporting the results of an investigation.

Secondary sources of information are a synthesis, a review or an interpretation of primary sources. Secondary sources include textbooks, biographies, documentaries, newspaper articles and websites.

Refer to the following two tables. The first shows a checklist of what to look for when assessing and selecting the best sources of information. The second shows how to set out the information that is required for the references section of your report in American Psychological Association (APA) seventh edition style. This is only one of several referencing systems that you might be required to use throughout your schooling and future career.

Selecting resources for the investigation	Tick ✔
The resource is:	
• credible and I can identify the author, author's expertise and publisher	
• current because the date of publication of the material is provided and is recent	
• factual and I know that it is objective material and not biased	
• accurate and all information is correct	
• relevant and covers the area I am investigating	
• readable and neither too simple nor too complex in its coverage of the material.	

Examples of information required for references and bibliographies (APA style)
Article in scientific magazine Author, initials. (year). Title of article. *Journal title volume number*(issue number), page numbers. Digital object identifier (DOI) or URL. Lee, M. L. (2017). Materials science: crystals aligned through graphene. *Nature 544*(7650), 301–302.
Book Author, initials. (year). *Title of book* (edition, if not first). Publisher. Chan, D., Commons, C., Commons, P., Derry, L., Freer, E., Huddart, E., Lennard, L., MacEoin, M., Moylan, M., O'Shea, P., Ross, R., & Vanderkruk, K. (2023). *Heinemann Chemistry 1* (6th ed.). Pearson Australia. Include names of all authors or editors up to 20. Special rules apply for 21 or more authors. You can find information about the rules online.
Internet Author, initials/name of organisation. (year). *Title of webpage or web document.* URL. Royal Society of Chemistry. (2017). *Periodic table.* http://www.rsc.org/periodic-table

Always remember to record the above information for each resource you use. It is important to accurately cite resources in your reference section as you use them, because it can be time-consuming and difficult to find this information later.

NOTE-TAKING AND ORGANISING NOTES

Note-taking and organising requires skill. Good note-taking helps you avoid plagiarism and provides excellent information to support your scientific report writing. Plagiarism is when you take someone else's ideas and words and present them as your own work. You plagiarise if you copy sections or sentences from sources or you cut and paste from the internet. It is acceptable to use the ideas of others but you must state clearly where the information has come from in your reference section.

The following table provides examples of original text, plagiarised text and acceptably rephrased text.

Original text	Plagiarised text	Rephrased text
Critical elements are elements that are heavily relied on for industry and society in areas such as energy, electronics and food production. The demand for critical elements is so high that current methods of obtaining or recovering the elements are often unsustainable. Future development of recycling methods will be crucial to allow their continued use.	Critical elements are heavily relied on for energy, electronics, food production and other areas of society. The high demand for critical elements means current methods of obtaining the elements are often unsustainable. Continued use will require future development of recycling methods.	Critical elements are important in society in many areas including production of energy, electronics and food. The way critical elements are currently obtained from mining or through recycling is not enough to supply current demand. Ways of improving recycling of critical elements are essential in the future.

There are various approaches to effective note-taking. Whatever technique you use, try to keep your notes brief and focus on key points. Examples include:

- dot point summaries
- underlining or highlighting text
- labelled diagrams
- flow charts—to show sequences
- concept maps—to show connections between ideas
- Venn diagrams—to show similarities and differences
- tables—useful for summarising longer and more complex information that has subparts and can incorporate any of the other note-taking techniques. Adapt the table to suit your style and the task.

The following (partially completed) table shows how this technique can be used to take notes for a secondary-sourced investigation.

Secondary-sourced investigation notes			
Many scientists believe that banning the sale of plastic water bottles and single-use plastics is necessary to reduce harm to our health, wildlife and the environment. Construct an argument for or against this statement.			
	Source 1 (e.g. book) Author: Year of publication: Title: Edition (if not first): Publisher's name:	Source 2 (e.g. internet) Author/organisation: Year: Title of webpage/document: URL:	Source 3 (e.g. science journal) Author: Date (year): Title of article: Journal title: Volume number: Issue number: Pages: DOI or URL (if electronic article):
Position of Australian states on banning single-use plastics		**Which states are leading the way on banning single-use plastics?** South Australia Western Australia Australian Capital Territory Queensland Tasmania Northern Territory New South Wales Victoria	
Impact of single-use plastics on the environment	• raw materials—crude oil, natural gas • disposal—land fill		

continues over

 ISBN 978 0 6557 0015 9

Secondary-sourced investigation notes – *continued*			
Reasons to ban single-use plastics	Increased demand for single-use plastics is unsustainable and causing environmental damage.		
Reasons not to ban single-use plastics but to limit their use			Single-use plastics are used in medical equipment, and health workers rely on them for hygiene and infection control.
Effects of single-use plastics on our health, wildlife and the environment	• Microplastics are being found in greater concentrations in rivers and oceans as well as in our food. These can cause harm. • Thousands of sea birds, sea turtles and seals are found dead every year after eating or becoming entangled in plastic.		

SCIENTIFIC WRITING

Scientists use a particular writing style. You should use this distinctive style to communicate your ideas. Scientific writing is:

- objective—it describes events rather than what people think or feel and is as free as possible of bias or personal opinion
- precise—it avoids exaggeration and uses qualified language
- formal—it uses scholarly language rather than colloquial or everyday language
- concise—it conveys information in short, clearly understandable sentences without unnecessary information
- simple—it uses short sentences where possible
- predominantly written in passive voice, although sometimes active voice can be used to avoid confusion. For example, 'The oxygen level was recorded hourly', in the passive voice is perhaps more difficult for a younger audience to understand than, 'We recorded the oxygen levels every hour'.
- structured to include headings, tables, diagrams and mathematical calculations.

Examples of unscientific and scientific writing are demonstrated in the following table.

Unscientific writing	Scientific writing
Subjective, biased: • The results were fantastic. • This produced a disgusting odour. • The breathtakingly beautiful bowerbird…	Objective, unbiased: • The results showed… • This produced a pungent odour. • The golden bowerbird…
Exaggerated: • The object weighed a huge amount. • The magnesium burst into huge flames. • Millions of ants swarmed over…	Accurate, precise: • The mass of the object was 250 kg. • The magnesium burned vigorously. • Ants swarmed over…
Everyday, informal language: • The bacteria passed away. • The results don't… • We guessed that… • Previous researchers were slack and missed…	Formal language: • The bacteria died. • The results do not… • It was hypothesised that… • Previous researchers have not found…

PRESENTING A REPORT ON A SCIENTIFIC INVESTIGATION

Scientific findings may be presented in a variety of ways. A common format for presenting at science conferences is a poster. Posters can get ideas across to a large audience in an organised, concise and creative way. Other common presentation formats are essays, reports, oral presentations and articles. Each presentation format has its own conventions. The following table summarises the characteristics of a number of presentation formats.

Presentation format	Characteristics/inclusions	
poster	• balance of text and visuals • title, subheadings • balanced layout • captions for figures and tables	• references • hierarchy of font size according to subheading level • consistent font style—no more than three fonts
report/article	• structured with an introduction, paragraphs and conclusion • includes subheadings	• mainly text • can include diagrams, graphs and tables
essay	• structured with an introduction, paragraphs and conclusion • introduction states focus of essay • each paragraph makes a new point supported by evidence	• each paragraph links back to last paragraph • a text-style presentation format—visuals at end in appendix • conclusion draws all ideas together but does not include any new information
oral presentation	• needs to be engaging • refer to cue cards but do not read from them • watch audience as you speak	• stand still and avoid fidgeting • look at audience and appear confident

PROOFREADING

After you have completed the investigation and prepared your presentation, it is important to think about and check what you have done.

Proofread your work to minimise errors and maximise effective communication of the ideas from your investigation. Use the following questions as a proofreading checklist.

Proofreading checklist	Tick ✔
Have I:	
• investigated the question fully?	
• expressed myself clearly to communicate my ideas well?	
• used the scientific writing style?	
• included data analysis?	
• checked spelling, punctuation and grammar?	
• included references?	
• met the requirements of the presentation format?	

ISBN 978 0 6557 0015 9

Study skills

There are a variety of techniques and strategies you can use to help you study. You may find you use different strategies in different situations. For example, you may prefer to highlight key phrases in your notebook throughout the year but make summaries of topics before an examination. The strategies you choose will depend on personal preference and may not be the same as those used by your classmates.

Effective study skills involve more than the learning strategies you use. Equally important is when you use those skills. It is more effective to apply study skills throughout the year, revising and consolidating your knowledge as you progress through the course, rather than doing a rushed cram just before the examination. Revise your work regularly. Being organised and setting up a study plan is key to reducing stress.

GETTING ORGANISED

To get yourself organised, try the following steps.

- Use a diary to write down all homework and assessment tasks as soon as you get them. Note due dates and what is required.
- Be specific about the tasks you need to do. Rather than writing 'do Chemistry', it is more effective to note things such as which questions to answer and which page to look at in your student book.
- Write a list of everything you need to do each day. Tick off or cross out items as you complete them.
- Break down larger tasks into smaller separate parts that are manageable.
- Make sure your lists and planners are realistic. Do not set yourself more tasks than you can actually do.

STUDY TECHNIQUES

Studying requires concentration. Remove any distractions and factor in some breaks. Allow a 10-minute break every hour. Vary your study technique depending on the content to be learned and your personal preference. Although you may have already found a study technique that works for you, consider the following options.

Study technique	Tips
Highlighting Isotopes are different forms of an element. They have <u>the same atomic number</u> but <u>different mass numbers</u>. This is because they have the same number of protons and a <u>different number of neutrons</u>. As protons and neutrons both have a mass of 1, isotopes have <u>different masses</u>.	Highlight or underline key points as you read your notes or text.
Summary notes *IONIC COMPOUNDS* *Common properties of ionic compounds:* • *brittle, which means they shatter when hit with a hammer* • *hard, which means they are resistant to scratching* • *high melting point* • *unable to conduct electricity in solid state* • *good electrical conductor in liquid or aqueous states*	• Create a list of key headings and add some dot points about each heading. • Write your own summary of the key ideas in each chapter. • Use headings and subheadings. • Underline key words and key phrases. • Use simple diagrams. • Remember, the most effective chapter summaries are clear, concise and uncluttered.
Diagrams lowest energy orbit where electron is normally found higher energy orbits	• Diagrams can be used as a summary of key concepts. • Diagrams are useful memory triggers. • Diagrams present a lot of information in a visual way, with minimal text.

continues over

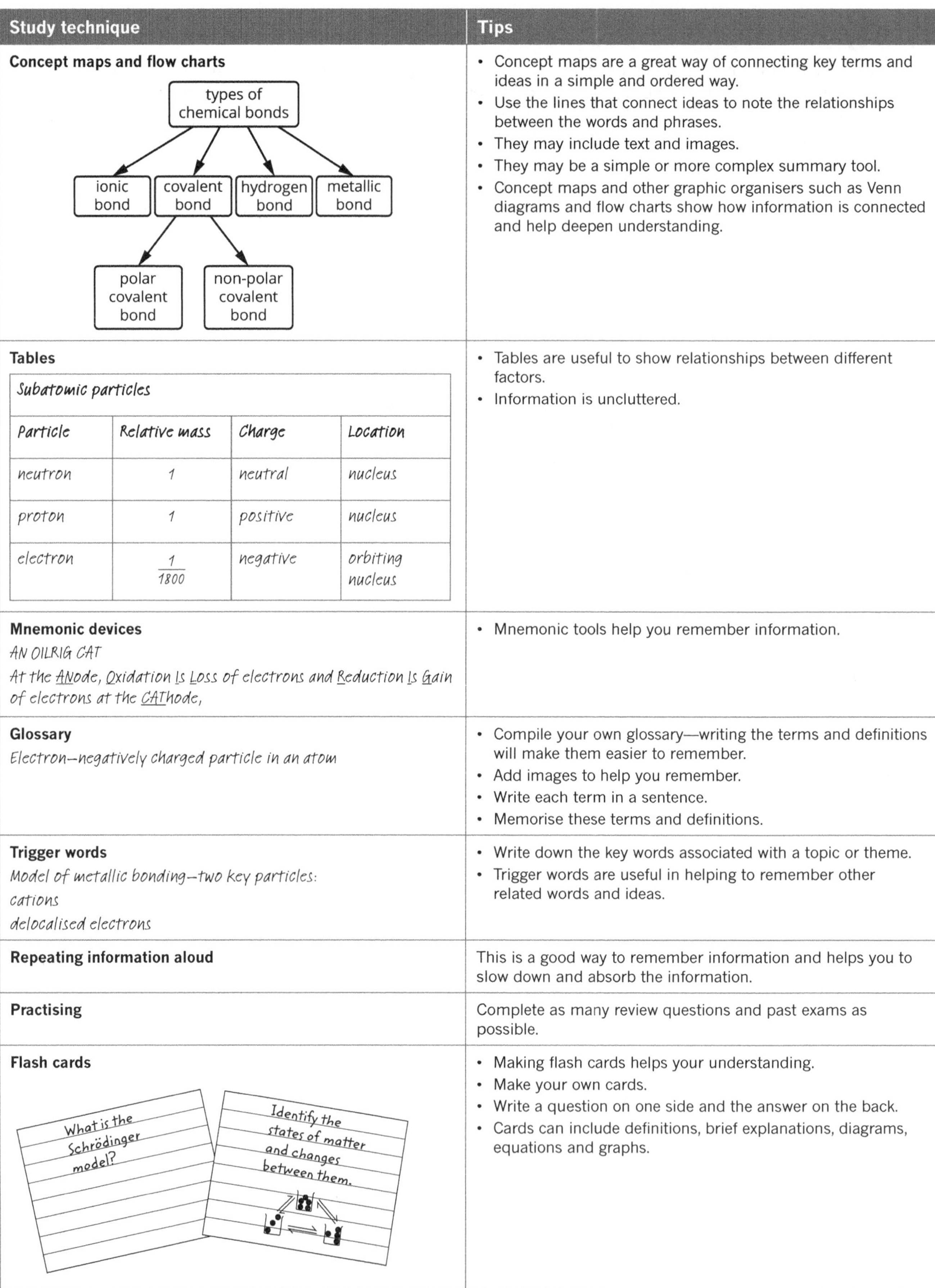

<table>
<tr><th>Study technique</th><th>Tips</th></tr>
<tr><td>Concept maps and flow charts
types of chemical bonds → ionic bond; covalent bond; hydrogen bond; metallic bond
covalent bond → polar covalent bond; non-polar covalent bond</td><td>• Concept maps are a great way of connecting key terms and ideas in a simple and ordered way.
• Use the lines that connect ideas to note the relationships between the words and phrases.
• They may include text and images.
• They may be a simple or more complex summary tool.
• Concept maps and other graphic organisers such as Venn diagrams and flow charts show how information is connected and help deepen understanding.</td></tr>
<tr><td>Tables
<table><tr><td colspan="4">Subatomic particles</td></tr><tr><td>Particle</td><td>Relative mass</td><td>Charge</td><td>Location</td></tr><tr><td>neutron</td><td>1</td><td>neutral</td><td>nucleus</td></tr><tr><td>proton</td><td>1</td><td>positive</td><td>nucleus</td></tr><tr><td>electron</td><td>$\frac{1}{1800}$</td><td>negative</td><td>orbiting nucleus</td></tr></table></td><td>• Tables are useful to show relationships between different factors.
• Information is uncluttered.</td></tr>
<tr><td>Mnemonic devices
AN OILRIG CAT
At the ANode, Oxidation Is Loss of electrons and Reduction Is Gain of electrons at the CAThode,</td><td>• Mnemonic tools help you remember information.</td></tr>
<tr><td>Glossary
Electron—negatively charged particle in an atom</td><td>• Compile your own glossary—writing the terms and definitions will make them easier to remember.
• Add images to help you remember.
• Write each term in a sentence.
• Memorise these terms and definitions.</td></tr>
<tr><td>Trigger words
Model of metallic bonding—two key particles:
cations
delocalised electrons</td><td>• Write down the key words associated with a topic or theme.
• Trigger words are useful in helping to remember other related words and ideas.</td></tr>
<tr><td>Repeating information aloud</td><td>This is a good way to remember information and helps you to slow down and absorb the information.</td></tr>
<tr><td>Practising</td><td>Complete as many review questions and past exams as possible.</td></tr>
<tr><td>Flash cards
What is the Schrödinger model?
Identify the states of matter and changes between them.</td><td>• Making flash cards helps your understanding.
• Make your own cards.
• Write a question on one side and the answer on the back.
• Cards can include definitions, brief explanations, diagrams, equations and graphs.</td></tr>
</table>

continues over

 ISBN 978 0 6557 0015 9

Study technique	Tips
Teaching someone	• Teach friends or family members. • Teaching a difficult concept to someone means you must first understand the concept yourself.
Handwriting notes	• Handwrite rather than type summary notes. • Remember, the examination requires you to handwrite answers. • Practise writing for long stretches of time and make sure your writing is legible.
Responding to feedback and self-correcting	• Check through all feedback from your teacher. • Highlight what was right or wrong. • Attempt to identify where you have errors and rework the answer to get it right.

Examination preparation

In the weeks before the examination, begin your exam preparation. The earlier you begin revising, the easier it will be. It is also helpful to begin practising exam-style questions as early as possible, not just in the weeks before the exam.

Like most skills, practice will improve your ability to do exams and to handle different types of exam questions. Doing practice exams is vital because you gain experience in:

- using reading time effectively
- allocating the right amount of time to each question
- working to a time limit
- reading and interpreting questions
- understanding what is required for each question
- planning answers
- deciding on relevant information
- proofreading/checking your answers
- writing efficiently for the duration of the exam.

Use the following checklist as a reminder of your study program.

Study program checklist	Tick ✔
Have I:	
• revised all areas of the course?	
• made a summary of the important points in each topic?	
• used effective study techniques to go over my revision notes?	
• looked at and worked through practice exam papers?	
• answered practice exam questions within the appropriate time limit?	

EXAMINATION STRATEGIES

Familiarise yourself with the conditions of the examination well before the day you sit the exam. You should know:

- the amount of reading time allowed in the exam before writing begins
- the amount of writing time allocated
- any particular equipment allowed and/or required, such as a calculator, pencils, pens and a ruler
- strategies to tackle the exam.

Exam strategies are listed in the following table.

Exam strategies
Reading time • You will be given reading time at the beginning of the exam (usually 15 minutes). • Remember that no writing at all is allowed during this time—no note-taking, no highlighting or underlining. • Read the instructions. • Read through the short-answer questions first—this will give you an overall sense of the themes of questions that require written responses. • Read the multiple-choice questions next. • Decide the order in which you will answer questions. Start with what you consider to be the easiest question to build confidence.
Writing time • For Units 1 and 2, your teacher will specify the length of your writing time. • Begin with the multiple-choice questions. • Answer every multiple-choice question, even if you can only make an educated guess. • If you are unsure of an answer to a multiple-choice question, mark it so you can come back to it if time allows. • Attempt the short-answer questions next. • Attempt the easiest short-answer questions first and work your way to the more challenging questions.
Tips for answering questions • Carefully read each question, underlining key words. • Be aware that most questions are structured so they become more challenging towards the end. You may not be able to answer the last part of a question, but you will have earned most of the marks by answering earlier parts. • Look carefully at any diagrams, pictures, tables and graphs and make sure you understand their relevance to the questions involved. • For questions with graphs, read the graph title and the labels on the axes carefully so that you can establish the relationship the graph is showing. • For questions with tables, read the headings on the columns and rows carefully so that you can analyse the content of the table effectively. • Check for key words in a question. Highlight them but do not colour the whole question. • Use correct spelling—it can mean the difference between scoring a point and not scoring a point. • Plan your answers before you write, remembering to address the exam criteria. • For questions with parts, read the whole question first. This gives you an overall picture of the question. It will also help to ensure that you do not repeat yourself in subsequent parts of the question. • Once you have answered the question, re-read your answer and then re-read the question to ensure that you have actually answered all of the question. • When writing a definition, avoid using the word you are defining in your definition. • If giving values from a graph, use a ruler to line up points with the axes so you can be accurate, and always include units in your answer. • Be sure to attempt all questions. • Read over your answers to pick up careless errors—the mind is faster than the hand and you may not always write what you intend (especially when you have limited time). • Write legibly. If the assessor can't read your answer, they can't mark it as correct. • Keep an eye on the time. • Never leave an exam early. Use any spare time to re-read and check your answers.
Exam cues • The number of marks allocated to a question provides a clue about how much you are expected to write. Two marks usually means you need to make a minimum of two points. • The number of lines allowed for the answer indicates the length of the expected answer. If your writing is large, you may need to turn the page over and continue on the back. Make sure you indicate to the assessor that they must turn to the back of the page for the rest of the answer. Where the answer continues, clearly state that it is the continuation of the question and state the question number.

ISBN 978 0 6557 0015 9

UNIT 1

How can the diversity of materials be explained?

AREA OF STUDY 1

How do the chemical structures of materials explain their properties and reactions?

Outcome

On completion of this unit the student should be able to explain how elements form carbon compounds, metallic lattices and ionic compounds, experimentally investigate and model the properties of different materials, and use chromatography to separate the components of mixtures.

Key knowledge

Elements and the periodic table

- the definitions of elements, isotopes and ions, including appropriate notation: atomic number; mass number; and number of protons, neutrons and electrons
- the periodic table as an organisational tool to identify patterns and trends in, and relationships between, the structures (including shell and subshell electronic configurations and atomic radii) and properties (including electronegativity, first ionisation energy, metallic and non-metallic character and reactivity) of elements
- critical elements (for example, helium, phosphorus, rare-earth elements and post-transition metals and metalloids) and the importance of recycling processes for element recovery

Covalent substances

- the use of Lewis (electron dot) structures, structural formulas and molecular formulas to model the following molecules: hydrogen, oxygen, chlorine, nitrogen, hydrogen chloride, carbon dioxide, water, ammonia, methane, ethane and ethene
- shapes of molecules (linear, bent, pyramidal, and tetrahedral, excluding bond angles) as determined by the repulsion of electron pairs according to valence shell electron pair repulsion (VSEPR) theory
- polar and non-polar character with reference to the shape of the molecule
- the relative strengths of intramolecular bonding (covalent bonding) and intermolecular forces (dispersion forces, dipole-dipole attraction and hydrogen bonding)
- physical properties of molecular substances (including melting points and boiling points and non-conduction of electricity) with reference to their structure and bonding
- the structure and bonding of diamond and graphite that explain their properties (including heat conductivity and electrical conductivity and hardness) and their suitability for diverse applications

Reactions of metals

- the common properties of metals (lustre, malleability, ductility, melting point, heat conductivity and electrical conductivity) with reference to the nature of metallic bonding and the existence of metallic crystals
- experimental determination of a reactivity series of metals based on their relative ability to undergo oxidation with water, acids and oxygen
- metal recycling as an example of a circular economy where metal is mined, refined, made into a product, used, disposed of via recycling and then reprocessed as the same original product or repurposed as a new product

Reactions of ionic compounds

- the common properties of ionic compounds (brittleness, hardness, melting point, difference in electrical conductivity in solid and molten liquid states), with reference to the nature of ionic bonding and crystal structure
- deduction of the formula and name of an ionic compound from its component ions, including polyatomic ions (NH_4^+, OH^-, NO_3^-, HCO_3^-, CO_3^{2-}, SO_4^{2-} and PO_4^{3-})
- the formation of ionic compounds through the transfer of electrons from metals to non-metals, and the writing of ionic compound formulas, including those containing polyatomic ions and transition metal ions
- the use of solubility tables to predict and identify precipitation reactions between ions in solution, represented by balanced full and ionic equations including the state symbols: (s), (l), (aq) and (g)

Separation and identification of the components in mixtures

- polar and non-polar character with reference to the solubility of polar solutes dissolving in polar solvents, and non-polar solutes dissolving in non-polar solvents
- experimental application of chromatography as a technique to determine the composition and purity of different types of substances, including calculation of R_f values

Elements and the periodic table

Chemistry is the study of the composition, properties and reactions of matter.

All matter is made of **atoms**. A substance made up of only one type of atom is called an **element** and a substance made up of two or more different types of atoms is called a **compound.**

Elements can be monatomic (existing as single atoms), exist as clusters of atoms called molecules or form into large networks called lattices or **giant molecules**. Compounds can be molecules or lattices. Molecules and lattices form the basis of all of the materials that are used in everyday life.

Matter can be classified as a pure substance or a mixture. A mixture can be further classified as homogeneous or heterogeneous (Figure 1.1.1).

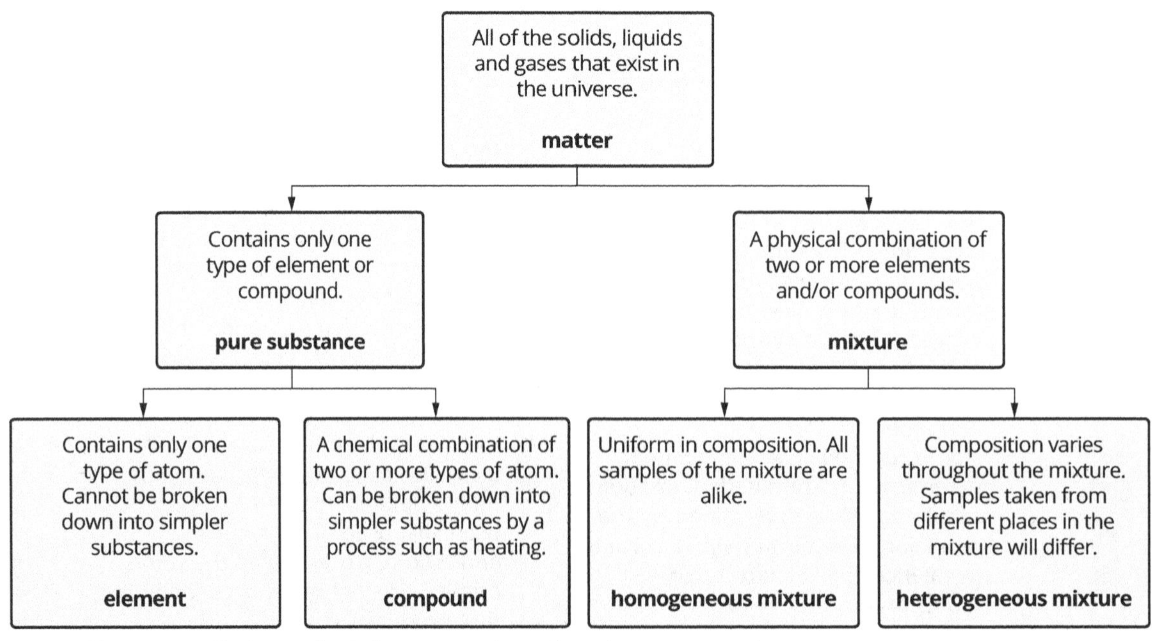

Figure 1.1.1 Matter is classified according to its components and how they are arranged.

 ISBN 978 0 6557 0015 9

KEY KNOWLEDGE

ATOMIC STRUCTURE

Each element is made up of one type of atom. Atoms contain three types of smaller subatomic particles called **neutrons, protons** and **electrons** (Table 1.1.1). The type of atom is determined by the number of protons in the nucleus. Atoms of the same type of element always contain the same number of protons.

Table 1.1.1 Subatomic particles

Particle	Relative mass	Charge	Location
neutron	1	neutral	nucleus
proton	1	positive	nucleus
electron	$\frac{1}{1800}$	negative	cloud surrounding the nucleus

The nucleus of an atom contains neutrons and protons held together by the nuclear strong force. The **atomic number** of an element indicates the number of protons in the nucleus and is given the symbol Z. The **mass number** of an element indicates the total number of nucleons (protons plus neutrons) (Figure 1.1.2).

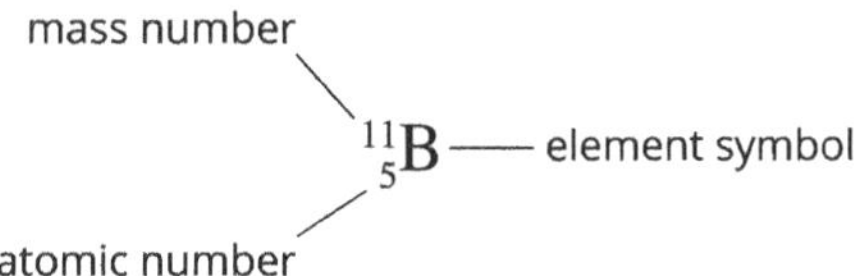

Figure 1.1.2 Representation of a boron atom

Isotopes

All atoms of the same element have identical atomic numbers. However, atoms of the same element can have different mass numbers; that is, they can have different numbers of neutrons. Atoms of the same element with different masses are called **isotopes** (Figure 1.1.3).

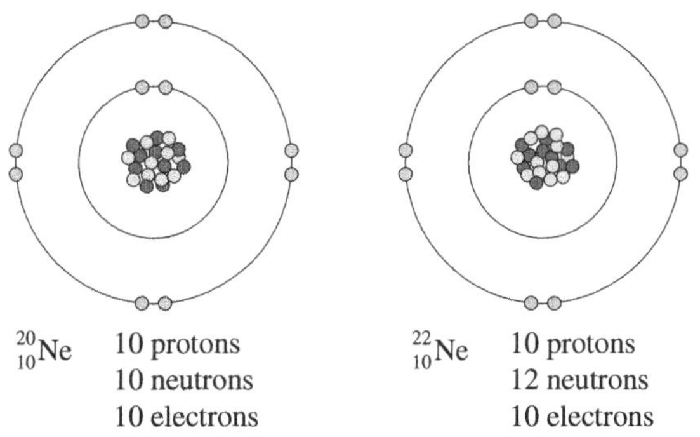

Figure 1.1.3 Isotopes of neon

ELECTRONIC STRUCTURE OF ATOMS

The Bohr model of the atom places electrons in certain well-defined orbits of fixed energies called **electron shells** (Figure 1.1.4).

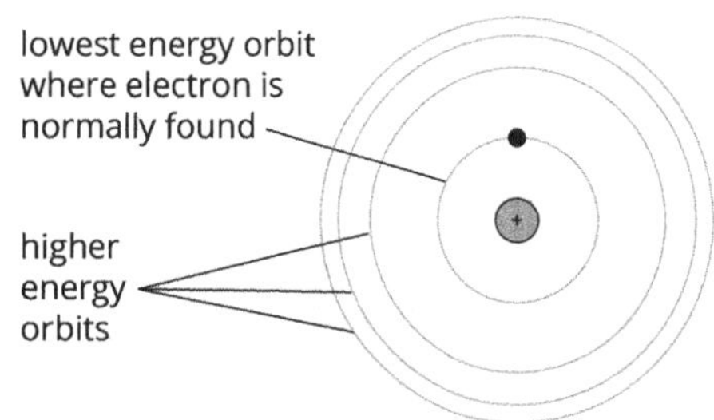

Figure 1.1.4 Bohr's shell model of the atom

Electrons in the same shell are a similar distance from the nucleus and have similar energy. Shells are numbered 1, 2, 3, 4, 5 and so on, from the nucleus outwards. A maximum number of electrons can occupy each shell (Table 1.1.2).

Table 1.1.2 Maximum number of electrons that can occupy each electron shell of an atom

Electron shell number (n)	Maximum number of electrons
1	2
2	8
3	18
4	32
n	$2n^2$

Electrons fill lower energy shells before higher energy shells. An **electronic configuration** for an element lists the number of electrons in each shell. For example, sodium has an atomic number of 11 and contains 11 electrons. Shells 1 and 2 are filled and shell 3 holds the 11th electron. The electronic configuration of sodium is 2,8,1.

The electron(s) in the outer shell of an atom are called valence electrons. Sodium has one valence electron. The valence electrons determine an element's chemical properties.

Subshell electronic configuration

Erwin Schrödinger refined Bohr's model by proposing that electrons should be regarded as having wave-like properties. In Schrödinger's model, electrons are not restricted to a given orbit but behave as negative clouds of charge in regions of space called **orbitals** (Figure 1.1.5).

Figure 1.1.5 An electron cloud around a nucleus

KEY KNOWLEDGE

Electrons make up most of the volume of an atom. Orbitals are located in **subshells** within shells (Table 1.1.3).

Table 1.1.3 Different levels of organisation of electrons

Level of organisation	Definition	Label used
shell	a major energy level within an atom	1, 2, 3, 4, 5 etc.
subshell	an energy level within a shell	*s*, *p*, *d*, *f*
orbital	a region in a subshell in which electrons move	

The Pauli exclusion principle states that each orbital may hold a maximum of 2 electrons (Figure 1.1.6).

- An *s* subshell has 1 orbital and can hold up to 2 electrons.
- A *p* subshell has 3 orbitals and can hold up to 6 electrons.
- A *d* subshell has 5 orbitals and can hold up to 10 electrons.
- An *f* subshell has 7 orbitals and can hold up to 14 electrons.

The subshells closest to the nucleus have the lowest energy and are filled first.

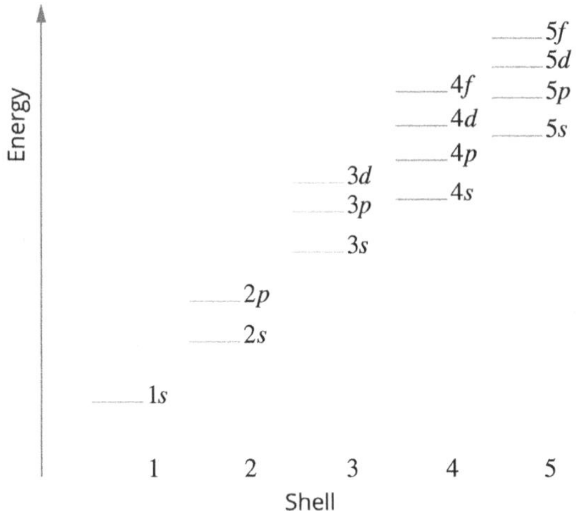

Figure 1.1.6 The energy levels in the atom

Note that some subshells are higher in energy than the subshells of the next shell (Figure 1.1.6). For instance, the unfilled 3*d* subshell is higher in energy than the unfilled 4*s* subshell. Therefore, the 4*s* subshell is filled before the 3*d* subshell.

Condensed electron configuration

The condensed electron configuration is an abbreviated version of the full configuration. The symbol of a noble gas, enclosed in square brackets, is used in the electron configuration to represent the inner-shell electrons. Only the valence electrons are shown in their shells or subshells.

The full and condensed electronic configurations of lithium, nitrogen, potassium and nickel are shown in Table 1.1.4.

Table 1.1.4 Electronic configurations of some elements

Element	Atomic number	Electronic configuration using subshell model	Condensed electronic configuration
lithium	3	$1s^2 2s^1$	[He] $2s^1$
nitrogen	7	$1s^2 2s^2 2p^3$	[He] $2s^2 2p^3$
potassium	19	$1s^2 2s^2 2p^6 3s^2 3p^6 4s^1$	[Ar] $4s^1$
nickel	28	$1s^2 2s^2 2p^6 3s^2 3p^6 3d^8 4s^2$	[Ar] $3d^8 4s^2$

Chromium and copper are exceptions to the usual order of subshell filling. In these cases, there is increased stability in having a half or completely full *d* subshell, so an electron from the 4*s* subshell is promoted to a 3*d* orbital. Their electronic configurations are:

Cr $1s^2 2s^2 2p^6 3s^2 3p^6 3d^5 4s^1$

Cu $1s^2 2s^2 2p^6 3s^2 3p^6 3d^{10} 4s^1$

Ions

Atoms are electrically neutral because they contain an equal number of negative electrons and positive protons.

Atoms can lose or gain valence electrons to form charged particles called ions. Ions can be positively or negatively charged, and have different-sized charges depending on the number of electrons lost or gained. Some examples are shown in Table 1.1.5.

Table 1.1.5 Charge on different ions

Ion	Number of protons	Number of electrons	Explanation for charge on ion
Na^+	11	10	one more proton than electron, so has a +1 charge
Mg^{2+}	12	10	two more protons than electrons, so has a +2 charge
Cl^-	17	18	one more electron than proton, so has a –1 charge
N^{3-}	7	10	three more electrons than protons, so has a –3 charge

- **You will now be able to complete Worksheets 1 and 2 and conduct Practical activity 1.**

ISBN 978 0 6557 0015 9

KEY KNOWLEDGE

THE PERIODIC TABLE

The periodic table is an extremely useful organisational tool for chemists. It can be used to identify patterns, trends and relationships between the structures and properties of elements.

The modern periodic table has the following features.

- Each box on the periodic table contains one element and information about that element (Figure 1.1.7).

Figure 1.1.7 The periodic table information for lithium. Different periodic tables may contain different information.

- Horizontal rows are called **periods**. Elements in the same period have the same number of occupied electron shells.
- Vertical columns are called **groups**. Elements in the same group have the same valence electron configuration.
- There are four large **blocks** of the periodic table (Table 1.1.6).

Table 1.1.6 Blocks in the periodic table

Block	Part of periodic table	Similarities in elements
s	groups 1 and 2	An *s* subshell is being progressively filled.
p	groups 13–18	A *p* subshell is being progressively filled.
d	transition metals groups 3–12	A *d* subshell is being progressively filled.
f	lanthanides and actinides	An *f* subshell is being progressively filled.

A full version of the periodic table can be found at the end of this book.

The position of an element on the periodic table provides a lot of information about the electron structure of its atoms. For example, knowing that an element is in group 2 and period 4 allows you to conclude that the atom has four occupied electron shells and two valence electrons.

Trends in the periodic table

Trends in properties observed in the periodic table (Table 1.1.7) reflect the changing numbers of protons and electrons.

Table 1.1.7 Trends in the periodic table

Measurement/ property	Description	Trend going down a group	Trend going from left to right across a period
atomic number	• a measure of the number of protons located in the nucleus of an atom	increases	successive elements increase by one
atomic radius	• a measure of the size of an atom, usually measured as distance from the nucleus to the valence electrons	increases	decreases
electronegativity	• the ability of an atom to attract shared electrons in a molecule • reflects how strongly the valence electrons are attracted to the nucleus	electronegativity decreases	electronegativity increases
first ionisation energy	• the minimum amount of energy required to remove the highest energy electron from an atom or ion • reflects how tightly the highest energy electron is held to the nucleus	first ionisation energy decreases	first ionisation energy increases
metallic/non-metallic character	• The more metallic an element, the more properties of a metal it has. • Metallic properties are often evident in elements whose atoms tend to readily lose their valence electrons.	metallic character increases	metallic character decreases
chemical reactivity of metals	• how readily a metal reacts with another element • reflects how readily a metal atom will release its valence electrons	reactivity of metals increases	reactivity of metals decreases
chemical reactivity of non-metals	• how readily a non-metal reacts with another element • reflects how readily a non-metal atom will accept an electron	reactivity of non-metals decreases	reactivity of non-metals increases, but last element in each period is an unreactive noble gas

- Going down a group, the number of occupied electron shells increases, so atomic radii are increasing and valence electrons are becoming further from the nucleus. This makes these electrons less tightly held so they can be more easily removed.
- Going across a period, the **effective nuclear charge** of successive elements increases by one (effective nuclear charge = number of protons − number of inner-shell electrons). As the effective nuclear charge increases, the valence electrons experience a greater attraction to the nucleus and are held more tightly. They become more difficult to remove.

CRITICAL ELEMENTS

Critical elements are elements that are heavily relied on for industry and society in areas such as energy, electronics and food production. The demand for critical elements is so high that current methods of obtaining or recovering the elements may be unsustainable. Future development of recycling methods will be crucial to allow their continued use. Some examples of critical elements are in Table 1.1.8.

Table 1.1.8 Examples of some critical elements

Element	Atomic number	Description	Reason for classification as critical	Major use
helium (He)	2	noble gas, second most abundant in universe, but rare on Earth	Helium is so light it easily escapes Earth's atmosphere. Atmospheric recovery is almost impossible.	super-cooling properties used in medical diagnostic equipment, and research into superconductors for applications including high-speed trains
phosphorus (P)	15	exists on Earth as phosphate rock	Mining reserves are in regions of the world difficult to access. Also large losses occur in the supply chain from mine to use on farms.	fertiliser for food production
indium (In)	49	post-transition metal, so called because it is located to the right of the transition metals	World supply depends on zinc mining. Zinc is itself endangered. Recovery of indium from scrap metal is currently very expensive.	LCD displays and photovoltaic panels
neodymium (Nd)	60	rare-earth element, so called because it is relatively rare in Earth's crust	Current recycling costs are high, and recovery is underdeveloped.	combines with other elements to make a strong, permanent magnet used in electric and hybrid vehicles as well as electronics such as loudspeakers, computer hard drives and mobile phones

- **You will now be able to complete Worksheet 3.**

Covalent substances

Covalent compounds consist of non-metal atoms covalently bonded to form molecules or lattices.

MOLECULAR SUBSTANCES

A non-metal atom generally has five, six or seven electrons in its outer shell. When non-metal atoms bond with other non-metal atoms, their relatively similar electronegativities result in them sharing their valence electrons. Discrete (individual) molecules are formed containing fixed numbers of atoms. Molecules then interact with other molecules to give the material its observable properties.

Intramolecular forces

In general, electrons are shared between non-metal atoms so that each atom has eight electrons in its outer shell (**octet rule**). A bond forms because the shared electrons are attracted to the positive nuclei of both atoms. Each pair of shared electrons constitutes a **covalent bond**, which is extremely strong and hard to break:

- One pair of shared electrons constitutes a single bond.
- Two pairs of shared electrons constitute a double bond.
- Three pairs of shared electrons constitute a triple bond.

The structures of molecules are most often represented by:

- **molecular formulas**, which give the number and type of atoms in each molecule

 ISBN 978 0 6557 0015 9

- **Lewis (electron dot) structures**, in which dots, crosses or both are used to represent each valence electron (Figure 1.1.8a)
- **structural formulas**, in which a single line represents each covalent bond. Non-bonding electron pairs are not shown (Figure 1.1.8b).

(a) H : O : H — Lewis structure

(b) structural formula

Figure 1.1.8 Different representations of a water molecule, which has the molecular formula H_2O: (a) a Lewis structure and (b) a structural formula

The four electron pairs of a stable outer shell form themselves into a tetrahedral shape. In this shape, the pairs of negatively charged electrons achieve maximum separation from each other within the atom according to valence shell electron pair repulsion (**VSEPR**) theory. The shape of the molecule is determined by the positions of atoms. Molecules can be **linear, bent, pyramidal** or **tetrahedral** (Table 1.1.9).

Polarity of molecules

Whether a molecule is polar or non-polar depends on the distribution of charge across the molecule. When two atoms bond that have different electronegativities, the atom with the higher electronegativity gains a greater share of the shared electron pair and thus carries a partial negative charge.

Table 1.1.9 Molecular shapes

Shape	Description	Example showing repulsion of all bonding and non-bonding electron pairs
linear	The two atoms of a diatomic molecular must be in a straight line.	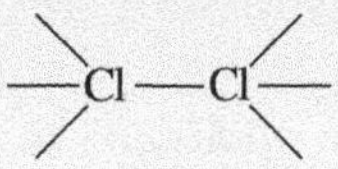 Cl—Cl chlorine
	The double bonds of triatomic molecules such as carbon dioxide repel each other—a straight line is as far apart as the electrons can get.	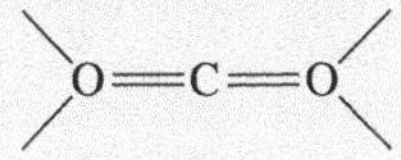 O=C=O carbon dioxide
bent	In a triatomic molecule, non-bonding pairs repel the bonding pairs of electrons. Because the bonding pairs also repel each other and try to achieve a linear shape, a bent shape results.	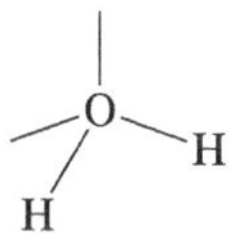 water (showing bonding and non-bonding electron pairs) 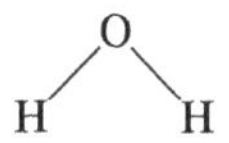 water (drawn without the non-bonding electron pairs, the bent shape of a water molecule is more obvious)
pyramidal	A molecule consisting of a central atom and three bonded atoms might be expected to form a flat triangular shape. Because these central atoms will usually have a non-bonded pair that repels the three bonded pairs, the triangular shape becomes a pyramid.	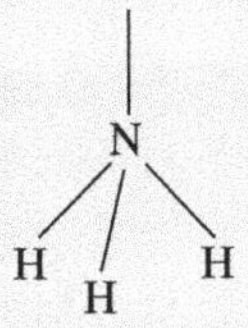 ammonia
tetrahedral	A molecule consisting of a central atom and four bonded atoms has four sets of bonding pairs trying to get as far apart as possible. The resulting molecular shape is described as tetrahedral.	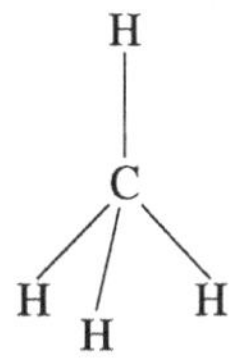 methane

The other atom carries a partial positive charge. The molecule is said to be a dipole and the bond is said to be a polar bond (Figure 1.1.9).

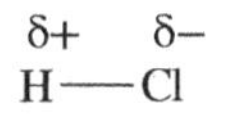

δ+ δ−
H—Cl

δ+ represents a small amount of positive charge.
δ− represents a small amount of negative charge.

Figure 1.1.9 A representation of a polar bond

A **polar** molecule:
- contains polar bonds
- has partial charges, distributed asymmetrically across the molecule (Figure 1.1.10).

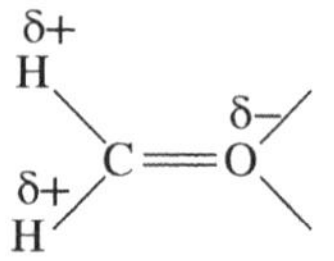

formaldehyde (or methanal)

Figure 1.1.10 The valence structure of formaldehyde, a polar molecule

A **non-polar** molecule may (Figure 1.1.11) or may not (Figure 1.1.12) contain polar bonds. If it contains polar bonds, the partial charges are distributed symmetrically across the molecule.

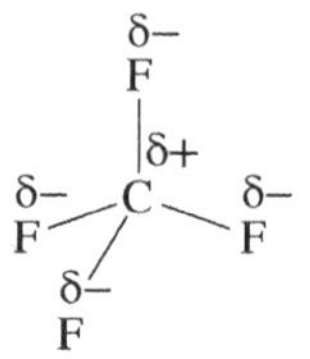

Figure 1.1.11 The C–F bond in the tetrafluoromethane molecule is polar, but the character of the molecule is non-polar.

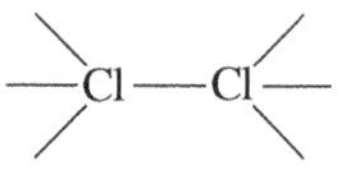

chlorine

Figure 1.1.12 The chlorine molecule is a non-polar molecule without polar bonds.

Intermolecular forces

The common properties of covalent molecular substances include the following.
- They have low melting and boiling points because of relatively weak intermolecular forces (a covalent molecular lattice can form if a covalent molecular substance is cooled enough for the molecules to form into a fixed lattice arrangement).
- They do not conduct electricity in solid or molten states because there are no free-moving particles present.

The most significant type of intermolecular force present depends on the polarity and structure of the molecule (Table 1.1.10).

The covalent bonding model contains a number of different types of bonds that have different relative strengths (Figure 1.1.13).

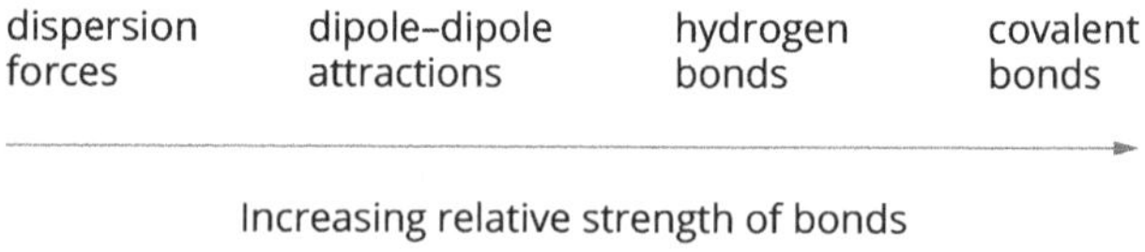

dispersion forces → dipole–dipole attractions → hydrogen bonds → covalent bonds

Increasing relative strength of bonds

Figure 1.1.13 Increasing relative strengths of different types of bonds involved in covalent bonding models.

- **You will now be able to complete Worksheets 4 and 5 and conduct Practical activity 2.**

CARBON LATTICES

Elemental carbon can form covalent lattice structures and other large molecules due to the carbon atoms bonding to each other in different ways. These different forms of an element are called allotropes. The allotropes of carbon have significantly different structures and properties.

There are different types of covalent lattices (Table 1.1.11).
- A **covalent network lattice** forms when covalent bonds extend throughout an entire substance. Diamond is a carbon covalent network lattice.
- A **covalent layer lattice** forms when covalent bonds extend throughout layers in a substance, and there are only weak dispersion forces between layers. Graphite is a carbon covalent layer lattice.

Table 1.1.10 Types of intermolecular forces

Force	When it is present	Bond occurs between	Relative strength
dispersion force	between all atoms/molecules, only attraction between non-polar molecules	two different atoms or molecules	weak, but increases as the size of the atom or molecule increases
dipole–dipole attraction	between polar molecules	a partially positive charge on one polar molecule and a partially negative charge on another	generally stronger than dispersion force but weaker than hydrogen bonds
hydrogen bond	between polar molecules that have an H bonded to an N, O or F	a partially positive H atom on one molecule and a lone pair of electrons on an N, O or F atom on another molecule	strongest intermolecular force but a lot weaker than a covalent bond

 ISBN 978 0 6557 0015 9

Table 1.1.11 Carbon lattices

Allotrope	Structure	Properties	Uses
diamond	Covalent network lattice. Each carbon atom is covalently bonded to four other carbon atoms in a tetrahedral arrangement.	• very hard • very high melting point • non-conductor of electricity	• jewellery • cutting tools • drills
graphite	Covalent layer lattice. Each carbon atoms is covalently bonded to three other carbon atoms in a layer. There is one delocalised electron per carbon atom. strong covalent bonds within layer weak dispersion forces between layers	• soft, greasy material • high melting point • good conductor of electricity	• electrodes • lubricant • pencils

● **You will now be able to conduct Practical activity 3.**

Reactions of metals

Common properties of metals include:

- high melting points
- lustrous or reflective when freshly cut or polished
- good conductors of heat
- good conductors of electricity
- malleable, which means they can be shaped by beating or rolling
- ductile, which means they can be drawn into a wire.

Not all metals have all of these properties. A limitation of the metallic bonding model is that it cannot explain these exceptions.

METALLIC BONDING MODEL

Metal atoms generally have one, two or three electrons in their outer shell and tend to lose these electrons in their interactions with other atoms.

The bonding model of metals (Table 1.1.12 on page 10) is a metallic lattice, which consists of a regular arrangement of stable, positively charged metal ions, called **cations**, surrounded by a 'sea' of freely moving delocalised electrons. The metal atoms achieve stability by releasing their valence electrons to become the surrounding 'sea'. Metallic bonds are an electrostatic attraction between the cations and the freely moving electrons.

Structure of metallic crystals

When a metal is cooled from its liquid form, or when a metal is produced from a solution by a chemical reaction, the lattice starts to form into crystals at many different places at once. Generally, the smaller the crystals, the less free movement there is of layers of ions over each other; hence, the metal becomes harder and less malleable. In addition, smaller crystals mean there are more disruptions to the lattice overall so the metal also becomes more brittle.

KEY KNOWLEDGE

Table 1.1.12 Summary of metallic bonding

Type of material	Constituent particles	Bonding model	Properties
Metal (e.g. iron)	• metal cations formed when metal atoms lose their valence electrons • delocalised electrons	positive ions 'sea' of delocalised electrons	• generally high melting temperature • good conductor of electricity • malleable • ductile • lustrous

REACTIVITY OF METALS

The **reactivity** of a metal is a reflection of its tendency to lose its valence electrons, a process called oxidation. Metals can be reacted with water, acid and/or oxygen in order to experimentally determine their ability to undergo oxidation.

- When a metal reacts with oxygen, it forms an oxide. Naturally occurring metals in Earth's crust often exist as oxides. Very unreactive metals, such as gold and platinum, exist in Earth's crust in their pure form.
- When a metal reacts with water, it produces hydrogen gas. The presence and vigour of the bubbling are an indication of the metal's reactivity.
- Metals that do not react with water may react with a dilute acid such as HCl. Hydrogen gas is a product, so the presence and vigour of bubbling are an indication of the metal's reactivity.
- Metals can also be reacted with other metal ions in solution in order to determine their relative reactivity. A more reactive metal will displace the ion of a less reactive metal from solution.

Reactivity series of metals

A **reactivity series of metals** (Figure 1.1.14) lists metals in order of their relative reactivity based on experimental data. The most reactive metal is at the top of the series and the least reactive is at the bottom.

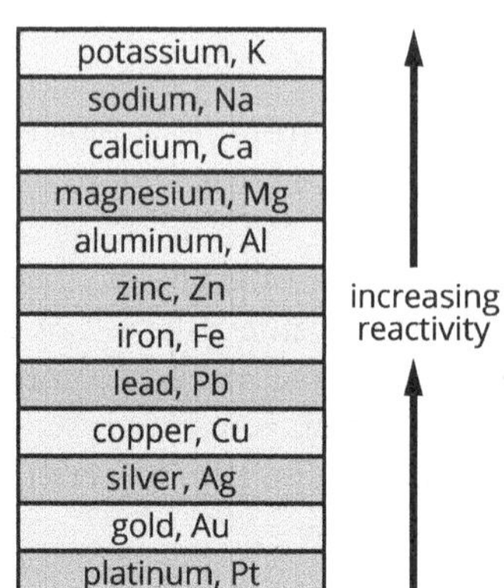

Figure 1.1.14 A reactivity series of metals lists metals from most reactive to least reactive.

Group 1 metals are the most reactive metals whereas transition metals are the least reactive.

A reactivity series is based on experimental observations. It has limitations because factors such as temperature, surface area of solids and concentration of reactants will also affect how vigorously a reaction occurs.

METAL RECYCLING

Metal recycling is a process of taking used metals and repurposing them to make new products. This supports **sustainability** through more responsible consumption and production of materials because it reduces:

- raw materials
- energy use
- air pollution
- water pollution
- waste.

Metals are generally suitable for recycling because they are strong, durable and malleable.

LINEAR AND CIRCULAR ECONOMIES

In a **linear economy**, raw materials are obtained from natural resources to make products that are discarded after use (Figure 1.1.15).

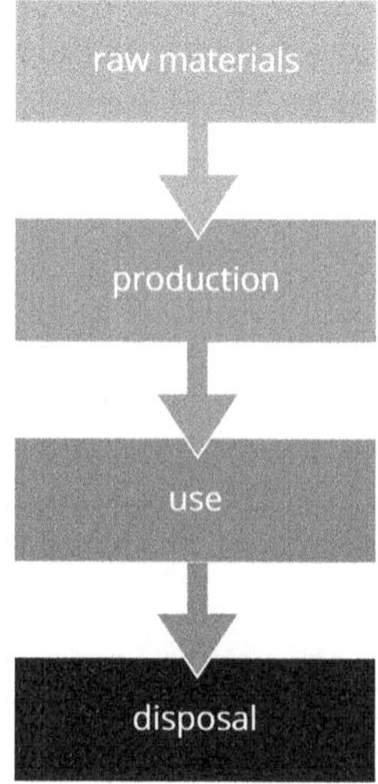

Figure 1.1.15 In a linear economy, there is no re-use or recycling of raw materials.

ISBN 978 0 6557 0015 9

In a **circular economy**, the use of natural resources is part of a continuous cycle (Figure 1.1.16). This supports more sustainable development because it uses available natural resources as efficiently as possible.

An example is in the production and recycling of a steel-based product. The raw material of iron is refined from iron ores mined from Earth. The iron is then used to produce steel, which is manufactured into useful products such as cars or used in construction. When the use of the steel ends (e.g. when a building is demolished), the steel, which is now scrap metal, is collected and repurposed to make new products. This reduces a reliance on natural resources for more raw materials and reduces waste going to land fill.

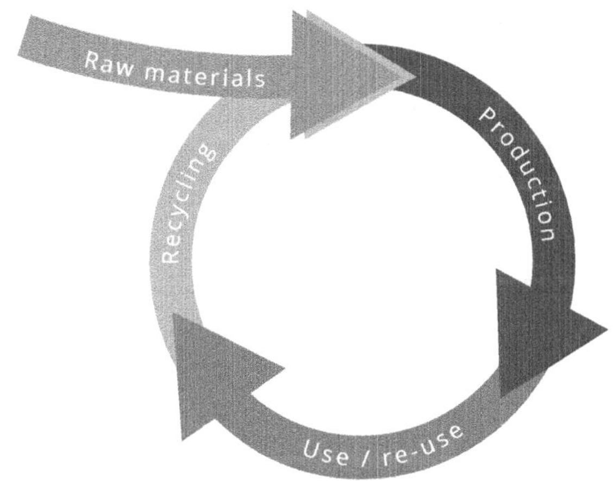

Figure 1.1.16 In a circular economy, products whose use has finished are re-used and recycled to reduce reliance on natural resources.

- **You will now be able to complete Worksheet 6 and conduct Practical activities 4 and 5.**

Reactions of ionic compounds

Common properties of ionic compounds include:

- brittle, which means they shatter when hit with a hammer
- hard, which means they are resistant to scratching
- high melting points
- cannot conduct electricity in the solid state
- good electrical conductors in molten (liquid) or aqueous states.

IONIC BONDING MODEL

The ionic bonding model (Table 1.1.13) is an ionic lattice that consists of a regular arrangement of positively charged metal **cations** and negatively charged non-metal ions called **anions**. The cations are formed when metal atoms achieve stability through transfer of their valence electrons to non-metal atoms, which then also become stable. Ionic bonds are the electrostatic attraction between the oppositely charged ions that holds the lattice together.

All models of bonding are only a representation of what is occurring at the atomic scale and so have limitations accordingly. For example, the diagram in Table 1.1.13 shows the space occupied by the anions and cations but does not show the actual ionic bonds.

IONIC CRYSTALS

Ionic compounds have a crystal structure, which means the ions are arranged in a regular geometric structure. The exact arrangement of ions in a crystal varies according to the size and ratio of the ions. Each crystal consists of a separate section of lattice that starts to form as a liquid is cooled, or water evaporates from a solution. Factors affecting the size and regularity of the crystals formed include the:

- rate of cooling
- rate of evaporation
- concentration of a solution
- purity of the solute.

Table 1.1.13 Summary of ionic lattices

Type of material	Constituent particles	Bonding model	Properties
ionic compounds (e.g. sodium chloride)	• cations, such as those formed when metal atoms transfer their outer shell electrons to non-metal atoms • anions, such as those formed when non-metal atoms gain electrons from metal atoms	+ cation – anion	• very high melting temperature • cannot conduct electricity when solid • can conduct electricity when molten or dissolved • brittle

KEY KNOWLEDGE

CHARGES ON IONS

The positive or negative charge on an ion indicates whether the atom has lost or gained electrons (Table 1.1.14). The size of the positive or negative charge indicates the number of electrons lost or gained. The charge on an ion is called its **electrovalency**.

Table 1.1.14 Charges on some common ions

Element	Number of electrons in outer shell	Tendency	Electrovalency	Ion
sodium	1	lose one electron	+1	Na^+
aluminium	3	lose three electrons	+3	Al^{3+}
oxygen	6	gain two electrons	–2	O^{2-}
chlorine	7	gain one electron	–1	Cl^-

IONIC FORMULAS

Anions and cations are produced by transfer of electrons from metal atoms to non-metal atoms during the formation of ionic compounds (Figure 1.1.17).

The ratio of metal atoms to non-metal atoms in a compound depends on the number of electrons being lost and gained by the atoms involved and is written in the ionic formula. The number of each ion is balanced so that the overall charge is neutral (Table 1.1.15).

Table 1.1.15 Formation of ionic compounds from ions of different electrovalencies

Cation	Anion	Formula	Name
Na^+	Cl^-	NaCl	sodium chloride
Mg^{2+}	Cl^-	$MgCl_2$	magnesium chloride
Ca^{2+}	F^-	CaF_2	calcium fluoride
Al^{3+}	O^{2-}	Al_2O_3	aluminium oxide

POLYATOMIC IONS

Some anions and cations are **polyatomic**, meaning the ion consists of more than one atom. Common polyatomic ions are:

- ammonium (NH_4^+)
- hydroxide (OH^-)
- nitrate (NO_3^-)
- sulfate (SO_4^{2-})
- hydrogen carbonate (HCO_3^-)
- carbonate (CO_3^{2-})
- phosphate (PO_4^{3-}).

Ionic formulas for compounds that contain polyatomic ions follow the same process of balancing overall charge. When more than one polyatomic ion is needed, the formula of the ion is written in brackets with the number of ions required written as a subscript after the brackets. (Table 1.1.16).

Table 1.1.16 Ionic compounds containing polyatomic ions

Name	Cation	Anion	Formula
magnesium nitrate	Mg^{2+}	NO_3^-	$Mg(NO_3)_2$
sodium sulfate	Na^+	SO_4^{2-}	Na_2SO_4
ammonium carbonate	NH_4^+	CO_3^{2-}	$(NH_4)_2CO_3$

TRANSITION METAL IONS

Many transition metals can form ions with different electrovalencies. The electrovalency of the transition metal ion is indicated by the roman numeral written after the cation in the name of the ionic compound (Table 1.1.17)

Table 1.1.17 Transition metal ions

Name	Constituent ions	Ionic formula
copper(I) chloride	Cu^+ and Cl^-	CuCl
copper(II) chloride	Cu^{2+} and Cl^-	$CuCl_2$
iron(II) chloride	Fe^{2+} and Cl^-	$FeCl_2$
iron(III) chloride	Fe^{3+} and Cl^-	$FeCl_3$

- **You will now be able to complete Worksheets 7 and 8.**

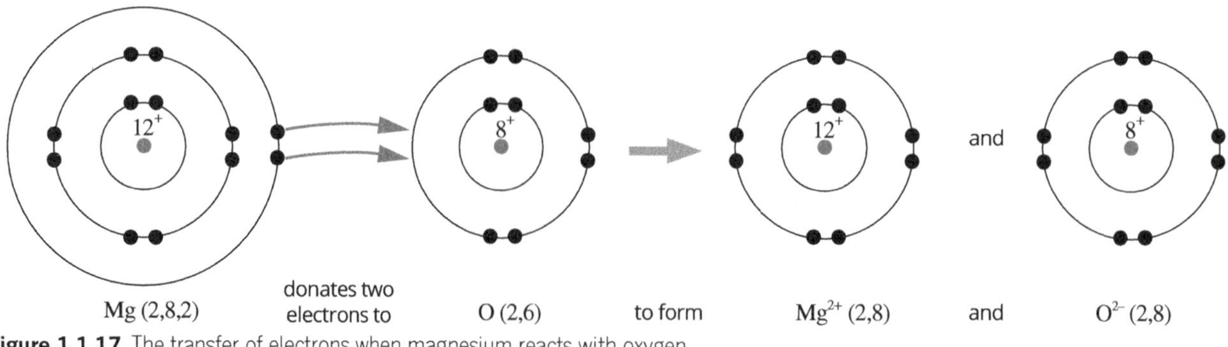

Figure 1.1.17 The transfer of electrons when magnesium reacts with oxygen

 ISBN 978 0 6557 0015 9

KEY KNOWLEDGE

SOLUBILITY TABLES

Some ionic compounds are highly soluble in water, whereas others are completely insoluble.

A solubility table (Tables 1.1.18 and 1.1.19) can be used to predict the solubility of an ionic compound in water.

Table 1.1.18 Relative solubilities of soluble ionic compounds

Soluble in water (>0.1 mol dissolves per L at 25°C)	Exceptions: insoluble (0.01 mol dissolves per L at 25°C)	Exceptions: slightly soluble (0.01–0.1 mol dissolves per L at 25°C)
most chlorides (Cl^-), bromides (Br^-) and iodides (I^-)	AgCl, AgBr, AgI, PbI_2	$PbCl_2$, $PbBr_2$
all nitrates (NO_3^-)	no exceptions	no exceptions
all ammonium (NH_4^+) salts	no exceptions	no exceptions
all sodium (Na^+) and potassium (K^+) salts	no exceptions	no exceptions
all ethanoates (CH_3COO^-)	no exceptions	no exceptions
most sulfates (SO_4^{2-})	$SrSO_4$, $BaSO_4$, $PbSO_4$	$CaSO_4$, Ag_2SO_4

Table 1.1.19 Relative solubilities of insoluble ionic compounds

Insoluble in water	Exceptions: soluble	Exceptions: slightly soluble
most hydroxides (OH^-)	NaOH, KOH, $Ba(OH)_2$, NH_4OH*, AgOH†	$Ca(OH)_2$, $Sr(OH)_2$
most carbonates (CO_3^{2-})	Na_2CO_3, K_2CO_3, $(NH_4)_2CO_3$	no exceptions
most phosphates (PO_4^{3-})	Na_3PO_4, K_3PO_4, $(NH_4)_3PO_4$	no exceptions
most sulfides (S^{2-})	Na_2S, K_2S, $(NH_4)_2S$	no exceptions

*NH_4OH does not exist in significant amounts in an ammonia solution. Ammonium and hydroxide ions readily combine to form ammonia and water.

†AgOH readily decomposes to form a precipitate of silver oxide and water.

PRECIPITATION REACTIONS

When two solutions are mixed, an insoluble substance called a **precipitate** sometimes forms (Figure 1.1.18).

The balanced chemical equation for a precipitation reaction has the general form:

ionic compound in solution (aq) + ionic compound in solution (aq) → precipitate (s) + ionic compound in solution (aq)

The symbols in brackets indicate the state of the reactant of product. In this equation:

- **s** is the symbol for solid
- **aq** is the symbol for solution, i.e. dissolved in water.

Other state symbols used in chemical equations are:

- **g** for gas
- **l** for liquid.

For example:

$$Pb(NO_3)_2(aq) + 2KI(aq) \rightarrow PbI_2(s) + 2KNO_3(aq)$$

In this case, PbI_2 is the precipitate. K^+ and NO_3^- ions remain in solution and are termed **spectator ions**.

An ionic equation omits spectator ions to give a more accurate picture of the reaction taking place.

For example:

$$Pb^{2+}(aq) + 2I^-(aq) \rightarrow PbI_2(s)$$

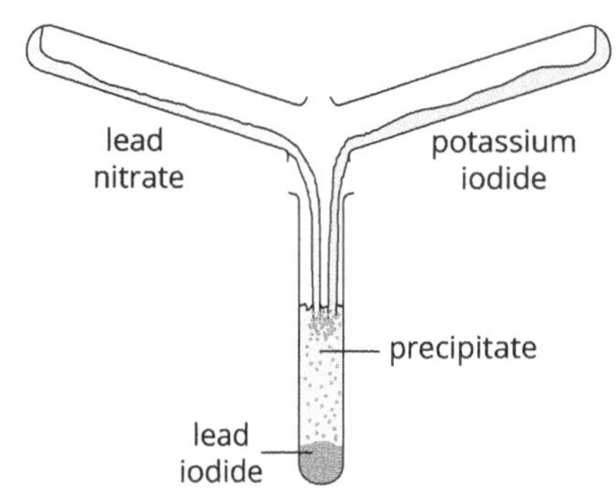

Figure 1.1.18 Mixing solutions of lead nitrate and potassium iodide forms a precipitate of lead iodide.

- **You will now be able to complete Worksheets 9 and 10 and conduct Practical activity 6.**

Separation and identification of the components in mixtures

When two or more elements or compounds are physically rather than chemically combined, it produces a mixture. The components in some mixtures can be separated and identified by a process called **chromatography**.

PRINCIPLES OF CHROMATOGRAPHY

To separate the components in a mixture of organic compounds, a sample of the mixture may be dissolved in a solvent. The solvent, or **mobile phase**, carries the sample as it passes over a solid material called the **stationary phase**. Separation of components occurs as a result of the components' different abilities to **adsorb** onto the stationary phase and **desorb** (dissolve) back into the mobile phase. The more strongly a component adsorbs to the stationary phase, the more slowly it moves over the stationary phase.

Influence of polarity

Polar substances tend to dissolve well in polar solutes, whereas non-polar solutes dissolve well in non-polar solvents. Polar substances tend to adsorb more strongly than non-polar substances do to polar stationary phases.

This means that changing the polarity of the solvent (mobile phase) and stationary phase in chromatography will produce different levels of separation of a mixture, based on the polarity of the components (Table 1.1.20).

Table 1.1.20 Effect of polarity in chromatography

Mobile phase	Stationary phase	Effect on components
non-polar	polar	The more polar components adsorb most strongly to the stationary phase and move the slowest. The less polar components remain more in the mobile phase and travel the fastest.

Simple chromatography techniques

Simple chromatographic techniques can be carried out in a laboratory with simple equipment. These include paper and thin-layer chromatography (TLC) (Table 1.1.21).

Components in a mixture can be identified by measuring their individual **retention factor, R_f** Retention factor is calculated by dividing the distance a component moves from the origin by the distance the solvent moves from origin. Components that have strongly adsorbed to the stationary phase will have a low retention factor.

For example, for the components in the TLC diagram in Table 1.1.21:

- $R_f(\text{component 1}) = \dfrac{4 \text{ cm}}{10 \text{ cm}} = 0.4$
- $R_f(\text{component 2}) = \dfrac{6 \text{ cm}}{10 \text{ cm}} = 0.6$

The values of R_f are characteristic for the component for the conditions under which the chromatogram was obtained.

● **You will now be able to complete Worksheets 11 and 12 and conduct Practical activity 7.**

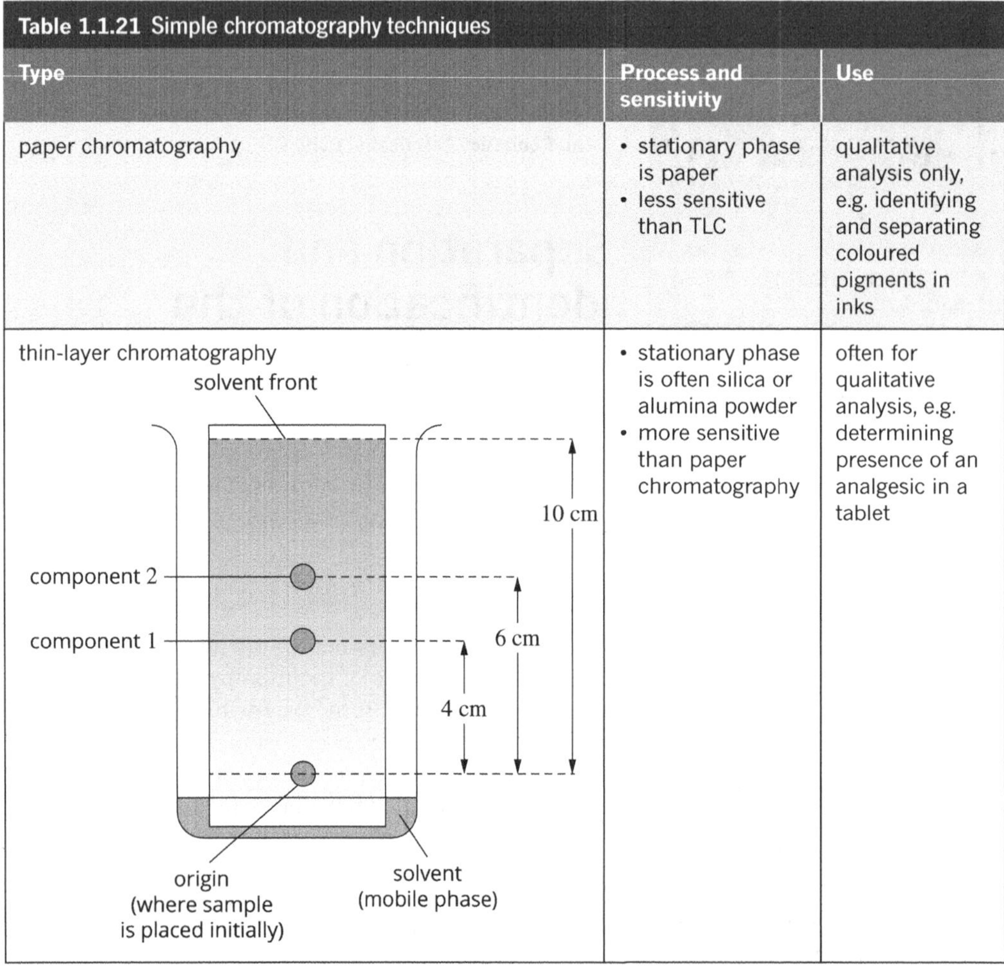

Table 1.1.21 Simple chromatography techniques

Type	Process and sensitivity	Use
paper chromatography	• stationary phase is paper • less sensitive than TLC	qualitative analysis only, e.g. identifying and separating coloured pigments in inks
thin-layer chromatography	• stationary phase is often silica or alumina powder • more sensitive than paper chromatography	often for qualitative analysis, e.g. determining presence of an analgesic in a tablet

 ISBN 978 0 6557 0015 9

Knowledge review—structure of the atom

Examine the three atoms represented by these shell models.

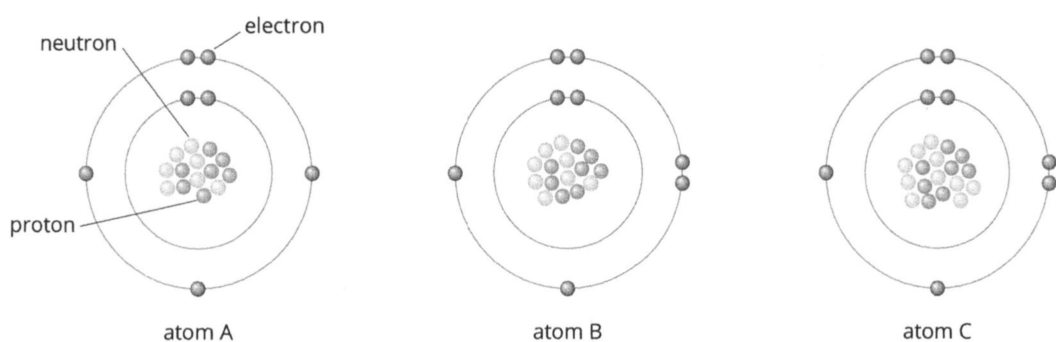

1 Complete the information for each atom.

	Atom A	Atom B	Atom C
Atomic number			
Mass number			
Name of element			

2 Which two atoms in Question 1 are isotopes? Explain.

3 The three types of subatomic particle are labelled in atom A.

a Which subatomic particle has a negative charge? ______________

b Which subatomic particle has a positive charge? ______________

c Which two subatomic particles are assigned a mass of 1? ______________________________

4 The diagrams above are representations of the structure of atoms. Describe two limitations of these diagrams.

5 When a chemical reaction occurs, the atoms in the reactants rearrange to become the products. Mark each statement below about chemical reactions as true or false.

Statement about chemical reactions	True or false?
Mass is always conserved in a chemical reaction.	
The rate of a chemical reaction will change with changes in temperature.	
The total mass of the products may be slightly less than or greater than the starting mass of the reactants.	
The number of protons in a particular atom might change during a chemical reaction.	
If there are six carbon atoms in the reactants of a chemical reaction, there must also be six carbon atoms in the products.	

ISBN 978 0 6557 0015 9

Writing electronic configurations—shells and subshells

1 The order of filling of subshells is represented in the first two columns in the table below. Fill in the number of electrons that will be present in each orbital type for each element shown. Carbon has been done for you as an example.

Shell	Subshell	Carbon (no. electrons = 6)	Chlorine (no. electrons =)	Copper (no. electrons =)	Vanadium (no. electrons =)	Selenium (no. electrons =)
1	*s*	2				
2	*s*	2				
	p	2				
3	*s*	0				
	p	0				
4	*s*	0				
3	*d*	0				
4	*p*	0				

2 Complete the following table for each element.

Element	Electron configuration using shells	Full electronic configuration using subshells	Condensed electronic configuration	Number of valence electrons
carbon	2,4	$1s^2 2s^2 2p^2$	$[He]2s^2 2p^2$	4
chlorine				
copper				
vanadium				
selenium				

3 Examine the following electronic configurations, both for the element fluorine:

- species A: $1s^2 2s^2 2p^5$
- species B: $1s^2 2s^2 2p^6$

Which species, A or B, represents a neutral fluorine atom? Give a reason for your answer.

4 Examine the following electronic configurations, both for the element sodium:

- species C: $1s^2 2s^2 2p^6 3s^1$
- species D: $1s^2 2s^2 2p^5 3p^2$

Which species, C or D, represents an excited sodium atom? Explain your answer.

 ISBN 978 0 6557 0015 9

Patterns in properties in the periodic table

1 Consider the elements lithium and fluorine. They are both located in period ______________. Lithium is in group ______________ and fluorine is in group ______________.

2 Consider the following properties of elements. Circle 'increases' or 'decreases' to describe the trend from left to right across the periodic table from lithium to fluorine.

Property	Trend
i atomic radius	increases/decreases
ii metallic character	increases/decreases
iii first ionisation energy	increases/decreases
iv electronegativity	increases/decreases

3 Add electrons to the atom outlines to represent shell models of the lithium and fluorine atoms. Also write the number of protons in each nucleus and the effective nuclear charge underneath the atoms.

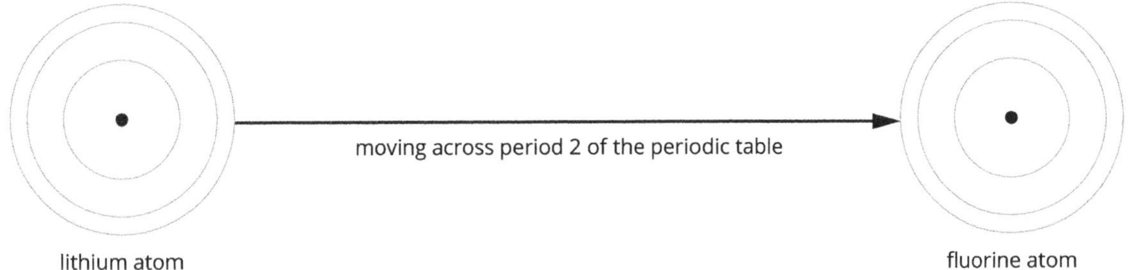

Number of protons: _______ Number of protons: _______

Effective nuclear charge: _______ Effective nuclear charge: _______

4 List any differences between the lithium and fluorine atoms in terms of their protons, electrons and effective nuclear charge.

5 Explain how these differences account for the different atomic radii of lithium and fluorine.

6 Explain how these differences account for the different metallic characters of lithium and fluorine.

Representations of molecules

Chemical bonds are a result of electrostatic attraction between different particles. Molecules form when electrons are shared between participating atoms, forming covalent bonds.

1 Identify the location of the positive and negative charges in the electrostatic attraction that causes a covalent bond.

negative charge: ______________________________

positive charge: ______________________________

2 Complete the table to identify the number of bonds each type of atom can form.

Element	Electron configuration (shells only)	Number of valence electrons	Number of covalent bonds the atom will form
carbon			
oxygen			
chlorine			
hydrogen			
nitrogen			

3 Use the information in the table in Question 2 to deduce the Lewis (electron dot) structure and structural formula of each molecule, then identify the numbers of bonding and lone pairs of electrons around each atom.

Molecular formula	Lewis structure	Structural formula	Number of bonding electrons in the molecule	Number of lone electrons around specified atoms
CH_4				C:
				H:
H_2O				H:
				O:
HCl				H:
				Cl:
NH_3				N:
				H:

4 Name the shape of each molecule in the table in Question 3.

CH_4 ______________________________

H_2O ______________________________

HCl ______________________________

NH_3 ______________________________

ISBN 978 0 6557 0015 9

Electronegativity and polarity of molecules

1 Choose from the terms listed to complete the summary statements outlining the key points about electronegativity and polarity in covalent molecules.

distance	partial	nuclei	polar	negative	shared
polarised	protons	non-polar	electronegativity	greater	

- The electrons in a covalent bond are ________________ by two atoms because they are attracted to the positive ________________ of both atoms.
- ________________ is the electron-attracting ability of an element. Different elements have different electronegativities according to the ________________ their valence electrons are from the nucleus and how many ________________ are in the nucleus. Patterns in the electronegativity of different elements are evident on the periodic table.
- In a covalent bond between atoms of elements with different electronegativities, the atom with the ________________ electronegativity pulls the shared pair closer to its side.
- The atom with the greater share of the negatively charged electrons gains a partial ________________ charge. The atom with a lesser share of the electron pair gains a ________________ positive charge.
- The bond is said to be ________________ due to the unequal sharing of electrons.
- A ________________ molecule results when the partial charges are not spread evenly across a molecule. A ________________ molecule has either no polar bonds or has polar bonds distributed symmetrically around the molecule.

2 Determine the polarity and type of intermolecular force present for each molecule in the table.

Molecule	Atom with highest electronegativity	Is the bond polarised? (yes or no)	Is the molecule polar? (yes or no)	Type of intermolecular force
H—Cl				
F–C(F)(F)F				
N(H)(H)H				
O=O				
H–O–H				
O=C=O				

3 Which two molecular substances would you expect to have the highest boiling temperature? Explain your answer.

____

____

ISBN 978 0 6557 0015 9

The metallic bonding model

1 A model of metallic bonding is shown. Add a definition or description of each label in the numbered spaces provided.

① Lattice:

____

____

③ Metal cation:

____

____

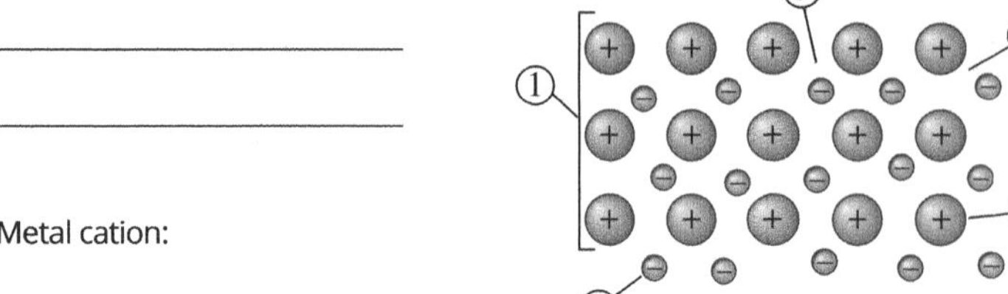

⑤ Metallic bond:

____

____

② Delocalised electron:

____

____

④ Electrostatic attraction:

____

____

2 Give one limitation of this representation of the metallic bonding model.

____

3 Next to each property of metals listed is a diagram. Add a written explanation of each property.

Lustrous

____

____

____

____

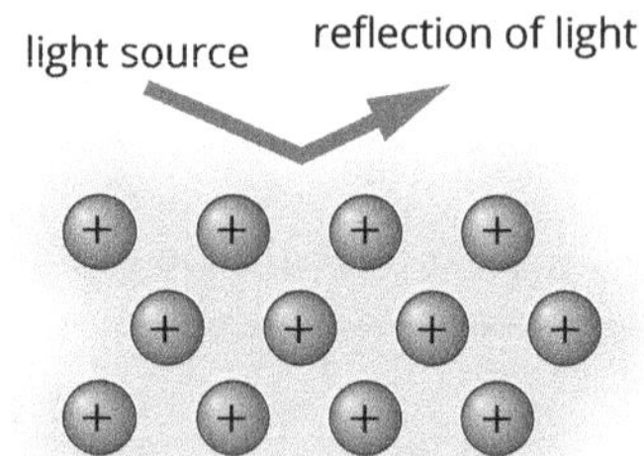

Malleable and ductile

____

____

____

____

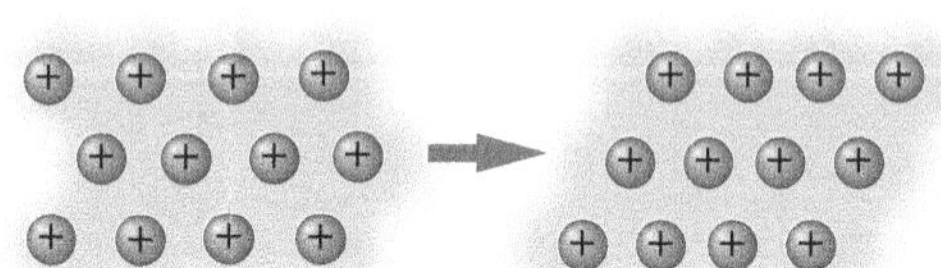

Good electrical conductor

____

____

____

____

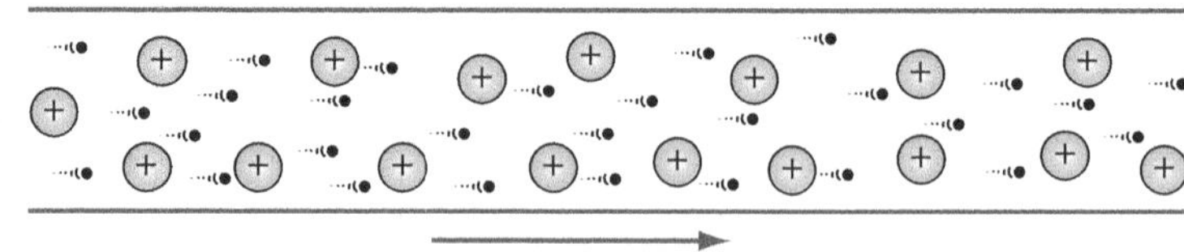

4 Use the metallic bonding model to explain why magnesium metal has a higher melting temperature than sodium metal.

____

____

____

 ISBN 978 0 6557 0015 9

WORKSHEET 7

The ionic bonding model

1 Label the following representation of the ionic bonding model.

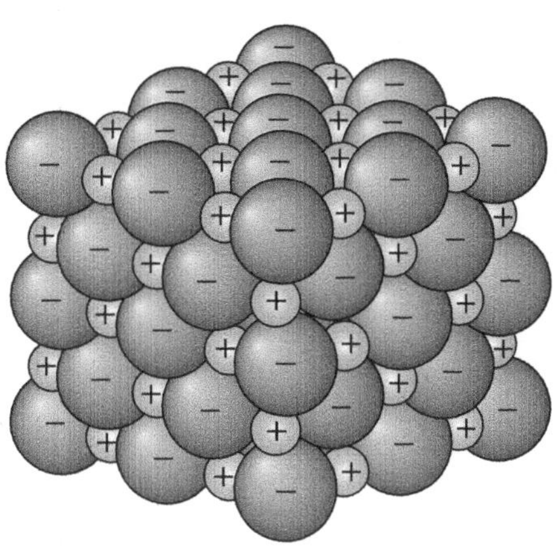

2 Each of the diagrams below demonstrates a property of an ionic substance. For each diagram:

- identify the property being demonstrated
- write an explanation of the property using the ionic bonding model.

Property: ______________________

Explanation

____

____

____

force

crystal shatters

Property: ______________________

Explanation

____

____

____

Property: ______________________

Explanation

____

____

____

ISBN 978 0 6557 0015 9

Writing ionic formulas

1 The following table gives a typical mineral analysis found on the label of a bottle of spring water.

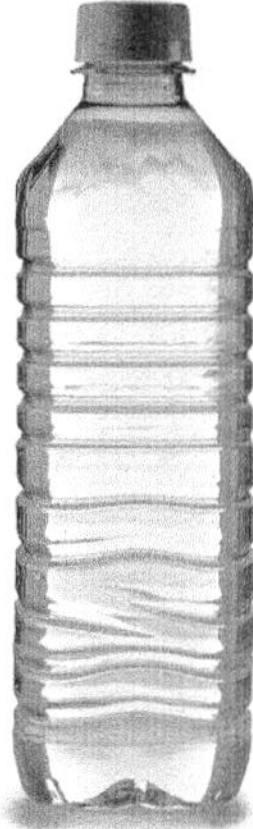

Ion	Concentration ($mg\,L^{-1}$)
bicarbonate	5
sodium	11
chloride	20
calcium	1
magnesium	4.7
potassium	0.5

Complete the following table.

Ion	No. electrons in outer shell of neutral atom	Charge on ion	Symbol of ion	Anion or cation?
sodium				
chloride				
calcium				
magnesium				
potassium				

2 With reference to electronic configurations, explain the following:

a Sodium and chlorine atoms form ions with opposite charges.

b The sodium ion has a charge of +1 and the magnesium ion has a charge of +2.

3 Another name for bicarbonate is hydrogencarbonate.

a Write the formula for the hydrogencarbonate ion. _______________

b The hydrogencarbonate ion is an example of a group of ions that contain more than one atom. These are termed _______________ ions.

4 List the names and chemical formulas of all possible compounds that could be formed from the ions listed on the label of the mineral water.

 ISBN 978 0 6557 0015 9

Solubility tables and predicting precipitation reactions

Table 1 Relative solubilities of soluble ionic compounds

Soluble in water (>0.1 mol dissolves per L at 25°C)	Exceptions: insoluble (<0.01 mol dissolves per L at 25°C)	Exceptions: slightly soluble (0.01–0.1 mol dissolves per L at 25°C)
most chlorides (Cl^-), bromides (Br^-) and iodides (I^-)	$AgCl$, $AgBr$, AgI, PbI_2	$PbCl_2$, $PbBr_2$
all nitrates (NO_3^-)	no exceptions	no exceptions
all ammonium (NH_4^+) salts	no exceptions	no exceptions
all sodium (Na^+) and potassium (K^+) salts	no exceptions	no exceptions
all ethanoates (CH_3COO^-)	no exceptions	no exceptions
most sulfates (SO_4^{2-})	$SrSO_4$, $BaSO_4$, $PbSO_4$	$CaSO_4$, Ag_2SO_4

Table 2 Relative solubilities of insoluble ionic compounds

Insoluble in water	Exceptions: soluble	Exceptions: slightly soluble
most hydroxides (OH^-)	$NaOH$, KOH, $Ba(OH)_2$, NH_4OH*, $AgOH$†	$Ca(OH)_2$, $Sr(OH)_2$
most carbonates (CO_3^{2-})	Na_2CO_3, K_2CO_3, $(NH_4)_2CO_3$	no exceptions
most phosphates (PO_4^{3-})	Na_3PO_4, K_3PO_4, $(NH_4)_3PO_4$	no exceptions
most sulfides (S^{2-})	Na_2S, K_2S, $(NH_4)_2S$	no exceptions

*NH_4OH does not exist in significant amounts in an ammonia solution. Ammonium and hydroxide ions readily combine to form ammonia and water.

†$AgOH$ readily decomposes to form a precipitate of silver oxide and water.

1 The following solutions are mixed together. Indicate whether or not you think a precipitate will form by placing a tick or cross in each box in the second column. For each predicted precipitate, write a full balanced chemical equation in the third column.

Solutions mixed	Precipitate? (✓ or ✗)	Balanced chemical equation
potassium chloride + silver nitrate		
copper(II) nitrate + sodium hydroxide		
magnesium nitrate + sodium chloride		
lead nitrate + potassium chloride		
sodium sulfate + calcium nitrate		

2 Outline an experimental method a student could follow to obtain a dry sample of calcium carbonate from solutions of calcium chloride and sodium carbonate.

ISBN 978 0 6557 0015 9

WORKSHEET 10

Writing full and ionic chemical equations

1 Choose from the terms listed to complete the summary statements outlining the key points about full and ionic chemical equations.

conserved	solid	aqueous	gaseous	atoms	rearrangement	
soluble	liquid	dissociate	ionic	created	reactant	full
destroyed	spectator	product	change	balanced	dissolved	

- Chemical equations are a way of representing the _______________ of atoms in a chemical reaction.
- Mass is always _____________ in a chemical reaction. Atoms are neither _____________ nor _____________.
- The number and types of atoms present before a chemical reaction are also present afterwards. Hence a chemical reaction must be _______________ for each type of atom.
- The state of each _______________ and _______________ is indicated by the following:
  - (s) indicates the substance is present in _______________ form
  - (aq) indicates the substance is _______________, i.e. _______________ in water
  - (l) indicates the substance is present in _______________ form
  - (g) indicates the substance is present in _______________ form.
- A _______________ chemical equation includes all of the _______________ present when a chemical reaction takes place. It does not always describe the chemical _______________ accurately because not all atoms present are always involved in chemical change. _______________ ionic compounds _______________ into their ions in solution. The ions that remain unchanged over the course of the reaction are termed _______________ ions.
- An _______________ chemical equation omits spectator ions. It focuses on the actual reaction taking place.

2 Identify the spectator ions in each reaction, and then write the ionic equation.

a $2NaBr(aq) + PbCl_2(aq) \rightarrow PbBr_2(s) + 2NaCl(aq)$

Spectator ions: _____________________

Ionic equation: ____

b $HCl(aq) + KOH(aq) \rightarrow KCl(aq) + H_2O(l)$

Spectator ions: _____________________

Ionic equation: ____

c $2HCl(aq) + MgO(s) \rightarrow MgCl_2(aq) + H_2O(l)$

Spectator ions: _____________________

Ionic equation: ____

 ISBN 978 0 6557 0015 9

WORKSHEET 11

Literacy review—comparing similar terms

In Area of Study 1, you learnt a number of new terms. Demonstrate your understanding of some of these terms by comparing the meaning of each pair of terms. Include examples wherever possible.

1 atomic number and mass number

2 electronegativity and electrostatic attraction

3 mobile phase and stationary phase

4 critical element and rare-earth element

5 electron configuration and condensed electron configuration

6 anion and cation

7 adsorption and desorption

8 ionic bond and covalent bond

9 polar molecule and non-polar molecule

ISBN 978 0 6557 0015 9

WORKSHEET 12

Reflection—How do the chemical structures of materials explain their properties and reactions?

The following table lists the key knowledge covered in this area of study.

1 Reflect on how well you understand the concepts listed. Rate your learning by shading the circle that corresponds to your current level of understanding for each one.

Key knowledge	Not confident ◄			► Very confident
Elements, isotopes and ions—notation, atomic number, mass number and numbers of protons, electrons and neutrons	○	○	○	○
Electron structure of atoms—electronic configuration, shells, subshells and orbitals	○	○	○	○
Periodic table—trends in physical and chemical properties of elements in periods and groups	○	○	○	○
Covalent bonding—shape and polarity of molecules and intermolecular forces	○	○	○	○
Chemical structures—metallic bonding and ionic bonding	○	○	○	○
Solubility of ionic substances and precipitation reactions	○	○	○	○
Chromatography—separation and identification of polar and non-polar components in a mixture	○	○	○	○

2 Consider the points you have shaded from 'Not confident' to 'Very confident'. List specific ideas you can identify that were challenging.

3 Write down two different strategies that you will apply to help further your understanding of these ideas.

ISBN 978 0 6557 0015 9

Using flame colours to identify elements

SUGGESTED DURATION

- 30 minutes

INTRODUCTION

Distinctive colours are obtained when certain metals or their salts are placed in a flame. Heating the metals in the flame gives the electrons in their atoms enough energy to move to higher energy levels. As the electrons return to lower energy levels, they give off the energy they had gained as a distinctive colour.

AIM

To observe some characteristic flame colours of metal ions and use the results to identify an unknown metal ion in solution

MATERIALS

- 100 mL spray bottles of each of the following 0.1 M solutions:
  - potassium chloride, $KCl(aq)$
  - calcium chloride, $CaCl_2(aq)$
  - barium chloride, $BaCl_2(aq)$
  - lithium chloride, $LiCl(aq)$
  - strontium chloride, $SrCl_2(aq)$
  - copper(II) chloride, $CuCl_2(aq)$
  - sodium chloride, $NaCl(aq)$
- 100 mL spray bottle containing a solution of an unknown metal chloride
- matches
- Bunsen burner and heatproof mat
- safety glasses/goggles
- disposable gloves

PRE-LAB SAFETY INFORMATION

Material used/other risks	Hazard	Control
0.1 M strontium chloride	• causes mild skin irritation • causes serious eye irritation	Wear safety glasses/goggles, disposable gloves and a lab coat.
0.1 M barium chloride	• may be harmful if swallowed	Wear safety glasses/goggles, disposable gloves and a lab coat.
0.1 M copper(II) chloride	• causes skin irritation • causes serious eye irritation • toxic to aquatic life with long-lasting effects	Wear safety glasses/goggles, disposable gloves and a lab coat.

Please indicate that you have understood the information in the safety table.

Name (print): ______________________

I understand the safety information (signature): ______________________

METHOD

1. Make the room is as dark as possible before starting this experiment. Tuck in the chairs and ensure there is a clear path to the exit in the event of an emergency. Put on your PPE, ensuring your lab coat is buttoned, your gloves are fitted correctly and comfortably, and your safety glasses/goggles are fitted over your eyes.
2. Set up your Bunsen burner on top of a heat mat. Light it and adjust it to the blue flame.
3. Select a spray bottle of one of the metal solutions and, holding the bottle approximately 10 cm from the top of the barrel of the Bunsen burner, spray a mist upwards into the Bunsen burner flame at a 45° angle. Spray away from other people, and towards a wall where possible.
4. Repeat this process for each of the remaining six solutions, waiting in between tests for the flame to return to a stable blue colour. Record the flame colours in Table 1.
5. Collect a spray bottle for the unknown solution and spray the solution in the Bunsen burner flame. Record the colour in Table 1.
6. Turn off your Bunsen burner, and spray and wipe down your bench area, alongside any walls or windows in the direction the solutions were sprayed. Return used equipment, wash equipment as necessary, and remove PPE as directed by your teacher.

RESULTS

Table 1 Flame colours of different metal chlorides

Type of metal in carbonate	Flame colour
potassium	
calcium	
barium	
strontium	
copper(II)	
lithium	
sodium	
unknown	

DISCUSSION

1 Identify which ion causes the flame colour: the metal cation or the chloride anion. Justify your answer.

2 Explain, in terms of the movement of electrons, why light is produced by these metals when they are heated in the Bunsen burner flame. Draw a labelled diagram as part of your answer.

ISBN 978 0 6557 0015 9

3 Explain how each metal produces a different flame colour.

4 Identify which metal ion you think was in the solid labelled 'unknown'. Describe the evidence you have used to make this choice.

5 a Identify the independent variable:

b Identify the dependent variable:

c Identify any controlled variables:

6 Identify at least two limitations of using a flame test to determine which metal ions are present in an unknown compound.

CONCLUSION

PRACTICAL ACTIVITY 2

Modelling

Making molecular models

SUGGESTED DURATION

- 50 minutes

MATERIALS

- commercial molecular model-building kit or golf-ball-size lumps of different colours of plasticine and toothpicks

INTRODUCTION

Molecular modelling is a useful way to represent covalent molecules in three dimensions and to relate this to two-dimensional representations such as Lewis (electron dot) structures and structural formulas.

AIM

To demonstrate the bonding and shape of a number of simple covalent molecules

METHOD

1 ▪ Before making any models, draw a Lewis (electron dot) structure of each molecule listed in Tables 1–3.
2 ▪ Using the appropriately coloured 'atoms', construct a model of each molecule.
3 ▪ Draw a three-dimensional representation of each molecule in the table.
4 ▪ Draw a structural formula for each molecule in the table.
5 ▪ Pack away model kits and return equipment.

RESULTS

Table 1 Single covalent bonds

Molecule	Diagram	Lewis structure	Structural formula
hydrogen, H_2			
chlorine, Cl_2			
hydrogen chloride, HCl			
water, H_2O			
ammonia, NH_3			
methane, CH_3			
ethane, C_2H_5			

ISBN 978 0 6557 0015 9

Table 2 Double covalent bonds			
Molecule	**Diagram**	**Lewis structure**	**Structural formula**
oxygen, O_2			
carbon dioxide, CO_2			
ethene, C_2H_4			

Table 3 Triple covalent bonds			
Molecule	**Diagram**	**Lewis structure**	**Structural formula**
nitrogen, N_2			

DISCUSSION

1 Consider each molecular model. Complete Table 4 by identifying each molecule's shape, whether it is polar or non-polar and the type of intermolecular force you would expect to exist between molecules for each of these substances.

Table 4 Molecule shape, polarity and intermolecular force			
Molecule	**Shape**	**Polarity**	**Strongest type of intermolecular force**
hydrogen, H_2			
chlorine, Cl_2			
hydrogen chloride, HCl			
water, H_2O			
ammonia, NH_3			
methane, CH_3			
ethane, C_2H_6			
oxygen, O_2			
carbon dioxide, CO_2			
ethene, C_2H_4			
nitrogen, N_2			

PRACTICAL ACTIVITY 2

2 Identify the polar molecules and explain why they are polar.

3 State at least two limitations of the models you have built in terms of how well they account for the properties of the compounds named in Question 2.

4 Based on the models you have made, identify one or more elements that can form double and/or triple bonds with its own atoms.

5 Based on the models you have made, identify two elements whose atoms never form double bonds.

6 Explain why the bonding pairs in CH_4 are arranged in a tetrahedral shape.

CONCLUSION

ISBN 978 0 6557 0015 9

PRACTICAL ACTIVITY 3

Controlled experiment

Comparing the physical properties of three covalent lattices

SUGGESTED DURATION

- 50 minutes

INTRODUCTION

Quartz (SiO_2), graphite (C) and dry ice (solid carbon dioxide, CO_2) are all composed of non-metal elements. In their solid forms, they represent different types of covalent lattice.

Silicon dioxide forms a covalent network lattice with silicon atoms covalently bonded to oxygen atoms in a 1 : 2 ratio throughout the lattice. Quartz, which is essentially pure SiO_2, is colourless and not normally used as a gem.

Graphite forms a covalent layer lattice in which carbon is covalently bonded to three other carbon atoms in hexagonal rings, forming layers.

Carbon dioxide consists of small non-polar covalent molecules. It is a gas at room temperature but exists in solid form at temperatures less than −78.5°C. Solid CO_2 (dry ice) exists as a covalent molecular lattice; that is, a lattice of molecules weakly bonded to each other. The dispersion forces between carbon dioxide molecules in solid carbon dioxide are so weak that dry ice sublimes (changes from a solid to a gas without passing through the liquid phase).

AIM

To compare the physical properties of a covalent network lattice, a covalent layer lattice and a covalent molecular substance

MATERIALS

- small samples of:
  - quartz, SiO_2
  - graphite, C
  - dry ice, CO_2 (to be conducted as a demonstration)
- hammer or mallet
- 3 evaporating dishes
- conductivity kit
- Bunsen burner
- sturdy bench mat or dissecting board
- tongs
- 3 × 100 mL glass beakers
- universal indicator and pH chart
- insulating gloves for handling dry ice
- safety glasses/goggles
- disposable gloves

PRE-LAB SAFETY INFORMATION		
Material used/other risks	**Hazard**	**Control**
dry ice (solid carbon dioxide)	• sublimes at −78.5°C • causes severe skin burns and eye damage • may cause drowsiness and dizziness	Use tongs and insulating gloves when handling dry ice. Wear safety glasses and a lab coat; cover dry ice with a cloth when crushing it. Teacher to conduct the testing as a demonstration to reduce student risk. (Note: pelletised dry ice is more suitable for this experiment.)

Please indicate that you have understood the information in the safety table.

Name (print): ______________________

I understand the safety information (signature): ______________________

METHOD

1. Put on your PPE, ensuring your safety glasses/goggles are fitted over your eyes.
2. Collect three lumps of quartz approximately 1 cm in size.
3. Collect three lengths of graphite approximately 1 cm in length. Old pencils with the outside wood removed, or pencil refills, are an excellent source of graphite.

EFFECT OF HEATING

4 ▪ Place the first sample of each substance on an evaporating dish. Observe any changes over the next 15 minutes.

5 ▪ After 15 minutes, heat the sample of quartz by holding it with tongs in a Bunsen flame. Observe whether the sample melts, chars or sublimes (solid changing directly to the gaseous state), or whether heating has no effect.

6 ▪ Repeat step 5 with the sample of graphite. Record your observations in Table 1.

COLOUR/TRANSPARENCY

7 ▪ Examine the second samples of quartz and graphite. Note their colour, ability to reflect light and transparency.

HARDNESS, BRITTLENESS, MALLEABILITY

8 ▪ Place the second samples, one at a time, on a sturdy bench mat. Lie this bench mat on the floor or on the ground outside.

9 ▪ Tap the sample with a hammer, gently at first, then with steadily increasing force. Test for hardness by observing whether the hammer marks the material. Test for brittleness by observing whether the material fractures. Test for malleability by observing whether it flattens out into a sheet.

ELECTRICAL CONDUCTIVITY

10 ▪ Take the third sample of each substance and use a simple conductivity kit to determine whether it conducts electricity.

SOLUBILITY

11 ▪ Place 60 mL of water in both beakers. Add a few drops of universal indicator to each beaker.

12 ▪ Add a small piece (about 0.5 cm^2) of each substance to the water in each beaker. Stir and allow it to stand for 5 minutes.

13 ▪ Observe whether the material dissolves in water. Note: If the substance dissolves in water, it can react further with the water to produce either a slightly acidic or a basic solution. This will be indicated by any changes in colour of the universal indicator.

14 ▪ Spray and wipe down your bench, return used equipment, wash equipment as necessary, and remove PPE as directed by your teacher.

DRY ICE TEACHER DEMONSTRATION

15 ▪ Observe your teacher conduct the five material tests on the dry ice and note your observations in Table 1.

RESULTS

Table 1 Physical properties of samples of quartz (SiO_2), graphite (C) and dry ice (CO_2)

Substance	quartz (SiO_2)	graphite (C)	dry ice (CO_2)
Melting properties			
Colour and transparency of solid form			
Hardness, brittleness and malleability			
Electrical conductivity			
Solubility			

 ISBN 978 0 6557 0015 9

DISCUSSION

1 Complete Table 2.

Table 2 Structure of the three carbon lattices

Substance	quartz (SiO_2)	graphite (C)	dry ice (CO_2)
Name of lattice type			
Diagram of lattice structure			

2 Identify the solid substance that was the hardest. Explain the reason for this with reference to the type of lattice.

3 Identify which substance conducted electricity, and explain why it was the only substance that could conduct electricity.

4 Explain the different melting properties of the three lattices.

CONCLUSION

Growing metal crystals

SUGGESTED DURATION

- 20 minutes (if warm agar is prepared before the lesson) plus 20 minutes to observe the crystals on the following 1 or 2 days

MATERIALS

- 0.5 g agar
- 40 mL deionised water
- approx. 10 mL of 0.1 M solution of one of the following: $AgNO_3$, $CuSO_4$, $SnCl_4$
- 1 cm × 4 cm strip of clean zinc sheet
- 250 mL glass beaker
- Petri dish
- hotplate or Bunsen burner and bench mat
- tripod
- gauze mat
- glass stirring rod
- scourer/sandpaper
- magnifying glass
- stereomicroscope (optional)
- safety glasses/goggles
- disposable gloves

INTRODUCTION

Zinc is a more reactive metal than silver, copper or tin. When a piece of zinc is placed in solutions that contain the positive ions of these other metals, the zinc metal reacts to form ions and the ion of the less reactive metal reacts to form its metal. The process can be summarised by an equation, such as the word and chemical equations for zinc + silver nitrate solution:

zinc + silver nitrate → zinc nitrate + silver

$$Zn(s) + 2AgNO_3(aq) \rightarrow Zn(NO_3)_2(aq) + 2Ag(s)$$

Each of the other compounds reacts in a similar way. These reactions are known as displacement reactions.

When a metal is formed quickly, as would be the case if the displacement reaction occurred in aqueous solution, the metal crystals tend to be small. It is difficult to see the details of their shape. However, if the formation of the metal is slowed down, as when the reaction occurs in agar, larger crystals form and it is easier to examine their shape.

AIM

To grow crystals of a range of different metals and to observe their shapes

PRE-LAB SAFETY INFORMATION

Material used/other risks	Hazard	Control
0.1 M silver nitrate, $AgNO_3(aq)$	• causes skin irritation • causes serious eye irritation • toxic to aquatic life with long-lasting effects • readily stains skin and clothing	Wear safety glasses/goggles, disposable gloves and a lab coat.
0.1 M copper(II) sulfate, $CuSO_4(aq)$	• causes mild skin irritation • toxic to aquatic life with long-lasting effects	Wear safety glasses/goggles, disposable gloves and a lab coat.
0.1 M tin(IV) chloride, $SnCl_4(aq)$	• causes serious eye irritation • causes skin irritation • may have long-lasting harmful effects on aquatic life	Wear safety glasses/goggles, disposable gloves and a lab coat.
Please indicate that you have understood the information in the safety table. Name (print): ____________ I understand the safety information (signature): ____________		

 ISBN 978 0 6557 0015 9

PRACTICAL ACTIVITY 4

METHOD

Note, agar solution could be prepared before class. Each group would then require 50 mL warm solution and begin at step 4.

1 ▪ Put on your PPE, ensuring your lab coat is buttoned, your gloves are fitted correctly and comfortably, and your safety glasses/goggles are fitted over your eyes.
2 ▪ Add 0.8 g of agar to 50 mL of deionised water in a beaker.
3 ▪ Warm gently, stirring until the agar is dissolved. Remove from the heat.
4 ▪ To the agar solution, add 10 mL of one of the 0.1 M $AgNO_3$, $CuSO_4$ or $SnCl_4$ solutions. Stir the mixture.
5 ▪ Pour the agar solution into a Petri dish.
6 ▪ Place the strip of zinc in the centre of the Petri dish. Do not move the Petri dish until the agar has set.
7 ▪ Once the agar has set, place a lid on the Petri dish to prevent the agar from drying out. Label your Petri dish with your group name and the sample name, using a marker or tape.
8 ▪ Repeat steps 4–7 with each of the metal salts.
9 ▪ Observe the formation of metal crystals over the next 1 or 2 days.
10 ▪ Spray and wipe down your bench, return used equipment, wash equipment as necessary, and remove PPE as directed by your teacher.

RESULTS

Type of metal crystal grown: ______________________________

Sketch the appearance of your metal crystal observed under a magnifying class. Also sketch the appearance of at least two other different metal crystals that have been grown in your class.

DISCUSSION

Write word and chemical equations for each chemical reaction that took place to produce a metal crystal.

CONCLUSION

PRACTICAL ACTIVITY 5

Controlled experiment

Reactivity of metals—student-designed practical activity

SUGGESTED DURATION

- 40 minutes + standing time

AIM

To compare the reactivity of a variety of metals in water, dilute acid and oxygen

PRE-LAB SAFETY INFORMATION		
Material used/other risks	**Hazard**	**Control**
1 M hydrochloric acid, HCl(aq)	• non-hazardous < 3 M	Wear safety glasses/goggles because of the reactions and products produced.
magnesium, ribbon	• when in contact with water, releases flammable gases, which may ignite spontaneously • burns with white-hot flame, which emits UV radiation and may cause eye damage	Wear safety glasses/goggles. Do not look directly at burning magnesium.
Please indicate that you have understood the information in the safety table. Name (print): I understand the safety information (signature):		

MATERIALS

- aluminium, Al; small turnings
- small strips of:
  - magnesium, Mg
  - iron, Fe
  - copper, Cu
  - zinc, Zn
- 1 M hydrochloric acid, HCl(aq)
- safety glasses/goggles

METHOD

In the space below, write a method for an experiment that will allow you to compare the reactivity of each of the metals available to you with cold water, hot water, 1 M hydrochloric acid and oxygen. When you have designed your method, write down the additional materials you require and have them both checked by your teacher before you proceed.

ADDITIONAL MATERIALS REQUIRED

ISBN 978 0 6557 0015 9

RESULTS

In the space below, draw up a results table to record all of your observations.

DISCUSSION

1 On the basis of your results, rank the metals in order of their reactivity from most reactive to least reactive.

2 Comment on the suitability of your method. What improvements could you make? How could you extend this investigation?

CONCLUSION

PRACTICAL ACTIVITY 6

Classification and identification

Precipitation reactions

SUGGESTED DURATION

- 50 minutes

INTRODUCTION

When two clear solutions containing dissolved ionic salts are mixed, an insoluble product called a precipitate may form and settle out of the mixture. Knowing which ions form precipitates is essential in many industrial processes, and in monitoring and maintaining natural waterways.

AIM

- To observe a number of reactions that involve the formation of a precipitate
- To distinguish between ions that form precipitates and those that are always soluble
- To use observations of patterns of solubility to identify unknown anions and cations in solution

MATERIALS

- dropper bottles, each containing a 0.1 M solution of one of the cations or a 0.5 M solution of one of the anions listed below. Solutions containing the desired cation will commonly be nitrates. Solutions containing the desired anion will commonly contain sodium or potassium ions as the cation.
  - Cations: Ba^{2+}, K^{+}, Mg^{2+}, Zn^{2+}, Cu^{2+}, Ca^{2+}, Ag^{+}
  - Anions: SO_4^{2-}, CO_3^{2-}, Cl^{-}, OH^{-}
- plastic well set or plastic grid
- marker pen
- 5 dropper bottles, randomly labelled A–E, containing solutions of unknown compounds
- safety glasses/goggles
- disposable gloves

PRE-LAB SAFETY INFORMATION

Material used/other risks	Hazard	Control
0.5 M sodium hydroxide, $NaOH(aq)$	• causes skin irritation • causes serious eye irritation	Wear safety glasses/goggles and a lab coat.
0.1 M calcium nitrate, $Ca(NO_3)_2(aq)$	• causes serious eye irritation	Wear safety glasses/goggles and a lab coat.
0.1 M copper(II) nitrate, $Cu(NO_3)_2(aq)$	• may be harmful if swallowed • toxic to aquatic life	Wear safety glasses/goggles and a lab coat.
0.1 M zinc nitrate, $Zn(NO_3)_2(aq)$	• causes skin irritation • very toxic to aquatic life with long-lasting effects	Wear safety glasses/goggles and a lab coat.
0.1 M silver nitrate, $AgNO_3(aq)$	• causes skin irritation • causes serious eye irritation • toxic to aquatic life with long-lasting effects • readily stains skin and clothing	Wear safety glasses/goggles, disposable gloves and a lab coat.
0.1 M potassium nitrate, $KNO_3(aq)$	• causes mild skin irritation	Wear safety glasses/goggles and a lab coat.
Please indicate that you have understood the information in the safety table. Name (print): I understand the safety information (signature):		

PART A • SOLUBILITY OF IONIC COMPOUNDS

METHOD

1 ▪ Put on your PPE, ensuring your lab coat is buttoned, your gloves are fitted correctly and comfortably, and your safety glasses/goggles are fitted over your eyes.

2 ▪ Using a marker pen, write the anions you are testing along the top of the plastic wells. Write the cations you are adding down the side of the wells. (Table 1 shows how your grid should look.) Place the well set on a dark background for easier observation.

3 ▪ Place two drops of the appropriate cation solution and one drop of the appropriate anion solution into each well, according to the labels you have written.

ISBN 978 0 6557 0015 9

4 ▪ If no precipitate is formed, record a cross.

5 ▪ If a precipitate forms, record 'ppt' (for precipitate) and record its colour. If the solution is cloudy, write 'cloudy' as your result.

6 ▪ Spray and wipe down your bench, return used equipment, wash equipment as necessary, and remove PPE as directed by your teacher.

RESULTS

Table 1 Observations of mixing of anions and cations

Cation/anion	Chloride, Cl^-	Sulfate, SO_4^{2-}	Hydroxide, OH^-	Carbonate, CO_3^{2-}
Mg^{2+}				
K^+				
Ca^{2+}				
Cu^{2+}				
Zn^{2+}				
Ag^+				
Ba^{2+}				

DISCUSSION

1 What generalisations can you make about the solubilities of:

a nitrates?

b chlorides?

c hydroxides?

d sulfates?

e carbonates?

2 Identify which cations:

a did not form any precipitates when they combined with the anions

b were insoluble with all of the anions tested.

3 What is the name of each precipitate that formed?

4 Write a net ionic equation to represent the formation of five of the precipitates observed.

PART B • IDENTIFYING UNKNOWN SOLUTIONS

METHOD

You have been given five unknown solutions A–E in dropper bottles. You need to identify the anion and the cation in each solution by designing tests based on your observations in Part A.

1 Outline a detailed method for identifying the anion present in solution A. You might like to present your method as a flow chart. Include necessary PPE and clean-up as part of your method.

 ISBN 978 0 6557 0015 9

2 Outline a detailed method for identifying the cation present in solution A. You might like to present your method as a flow chart. Include necessary PPE and clean-up as part of your method.

3 Refer to the Results section before conducting the practical. After your method has been approved by your teacher, conduct the experiment according to your method.

RESULTS

Construct a suitable results table for recording your results for the anion and cation tests for solution A.

DISCUSSION

1 Summarise your results by listing the cations and anions in the unknown solutions in Table 2 and writing the formula of each compound.

Table 2 Cations and anions in the unknown solutions

Solution	Anion	Cation	Formula of compound
A			
B			
C			
D			
E			

2 It is often possible to convert one salt solution to another by exploiting solubility differences. Briefly describe, in a series of steps, how you could convert a solution of calcium chloride to a solution of potassium chloride.

3 Evaluate whether the experiment provided an effective and efficient method of determining the positive and negative ions present in a series of unknown samples.

CONCLUSION

 ISBN 978 0 6557 0015 9

PRACTICAL ACTIVITY 7

Chromatography of a vegetable extract

SUGGESTED DURATION

- 50 minutes

INTRODUCTION

This experiment uses thin-layer chromatography (TLC) to separate the components in a green vegetable. TLC is a relatively simple form of chromatography that is similar to paper chromatography. A plate of silica or alumina is used as the stationary phase.

The separation of the components in a mixture by chromatography depends on their attraction to the stationary phase and their solubility in the mobile phase.

AIM

To separate the naturally occurring pigments in a green vegetable by thin-layer chromatography and determine the R_f values of the visible pigments

MATERIALS

- 20 g finely chopped portion of spinach or silverbeet leaves (frozen spinach works well)
- 2 mL cyclohexane, C_6H_6
- 12 mL propanone (acetone)
- 150 mL beaker (to place thin-layer chromatography plate in)
- fine sand
- 600 mL glass beaker
- 10 mL glass measuring cylinder
- mortar and pestle
- 4 cm × 6.5 cm thin-layer chromatography plate
- pencil and ruler
- spatula
- glass stirring rod
- glass capillary (melting point tube)
- hair dryer (optional)
- safety glasses/goggles
- disposable gloves

PRE-LAB SAFETY INFORMATION

Material used/other risks	Hazard	Control
propanone	• highly flammable liquid and vapour • causes serious eye irritation • may cause drowsiness or dizziness	Work in a fumehood. Keep away from flames and avoid breathing vapours. Wear safety glasses/goggles, disposable gloves and a lab coat.
cyclohexane	• highly flammable liquid and vapour • may be fatal if swallowed and enters airways • causes skin irritation • may cause drowsiness or dizziness • very toxic to aquatic life with long-lasting effects	Work in a fumehood. Keep away from flames and avoid breathing vapours. Wear safety glasses/goggles, disposable gloves and a lab coat.

Please indicate that you have understood the information in the safety table.

Name (print): ____________________

I understand the safety information (signature): ____________________

METHOD

1. Put on your PPE, ensuring your lab coat is buttoned, your gloves are fitted correctly and comfortably, and your safety glasses/goggles are fitted over your eyes.
2. Using a pencil, draw a line across a thin-layer chromatography (TLC) plate, 1 cm from the narrow end. Do not press too heavily or you will damage the surface of the plate.
3. Extract the pigment from a spinach or silverbeet leaf by grinding the chopped leaf with about 10 mL of propanone (acetone), using a mortar and pestle. Adding about half a spatula of fine sand will aid the crushing process.
4. In a fumehood, pour the propanone solution into a 100 mL beaker. Dip a capillary into the propanone mixture and use the solution that rises into the capillary to draw several very small spots on the pencil line on the chromatography plate, about 1 cm apart.
5. Allow the line to dry naturally, or dry it with a hair dryer, and repeat the procedure until very deep green spots are formed.
6. Use a measuring cylinder to pour 2 mL of propanone and 2 mL of cyclohexane into a 150 mL beaker. Use a stirring rod to mix the two liquids (they will not dissolve in each other completely).

7 ▪ Place the chromatography plate in the beaker so that the bottom of it is in the solvent and the plate rests against the rim of the beaker. The top of the solvent must be below the green spots on the plate. Invert a 600 mL beaker over the top of the smaller beaker.

8 ▪ Allow the solvent to rise up the plate to a height of about 5 cm.

9 ▪ Remove the plate, allow it to dry and use a pencil to mark the level the solvent has reached and the position of each component. (The colour of some components fades after several hours.) Complete the results table.

10 ▪ Spray and wipe down your bench, return used equipment, wash equipment as necessary, and remove PPE as directed by your teacher.

RESULTS

Sketch or take a photograph of the chromatogram you produced and stick into the space below. Label the distances travelled by the solvent front and by each component from the origin.

Distance of solvent front from origin:______cm

For each major component in the vegetable extract, describe its colour and calculate and record its R_f value in Table 1.

Table 1 R_f values of the components

Colour of band	Distance from origin (cm)	R_f value

DISCUSSION

1 Explain why the spots of plant extract placed on the plate should be as small as possible.

2 The orange colour of carrots and pumpkins is due to the presence of the chemical carotene. Describe how you could identify whether one of the compounds found in the vegetable extract analysed in this experiment is carotene.

 ISBN 978 0 6557 0015 9

3 Explain why different solvents produce different chromatograms.

4 Explain why the line where the extract is placed on the plate must be above the level of the solvent at the beginning of the experiment.

5 Explain why you made several spots of the same plant extract on the plate, rather than just one.

6 Compare the R_f values you calculated using your chromatogram with the values calculated by other groups in your class. Comment on whether the results appear to be reproducible and what that indicates about the validity of this practical activity.

7 If you were to carry out this activity again, identify steps you might do differently in order to improve your results.

CONCLUSION

EXAM-STYLE QUESTIONS

Multiple-choice questions

Question 1

Identify the particles in the isotope of iodine that is represented by the symbol $^{131}_{53}I$.

A. The isotope has 53 protons and 78 neutrons.

B. The isotope has 53 neutrons and 78 protons.

C. The isotope has 53 neutrons and 131 protons.

D. The isotope has 53 protons and 131 neutrons.

Question 2

Which of the following electronic configurations could be for a $^{52}_{24}Cr$ atom in an excited state?

A. $1s^2 2s^2 2p^6 3s^2 3p^6 3d^5 4s^2$

B. $1s^2 2s^2 2p^6 3s^2 3p^6 3d^5 4s^1$

C. $1s^2 2s^2 2p^6 3s^2 3p^6 3d^2 4s^2$

D. $1s^2 2s^2 2p^6 3s^2 3p^6 3d^5 4p^1$

Question 3

Identify the correct formula of the ionic compound made from strontium and chlorine.

A. SrCl

B. $SrCl_2$

C. Sr_2Cl

D. Sr_2Cl_2

Question 4

Identify which of the following molecules is polar.

A. O_2

B. CH_4

C. CO_2

D. NH_3

Question 5

Select the number of lone electron pairs in a molecule of methane, CH_4.

A. 0

B. 1

C. 2

D. 4

Question 6

Select the best description of the shape of a molecule of methane, CH_4.

A. bent

B. linear

C. pyramidal

D. tetrahedral

ISBN 978 0 6557 0015 9

EXAM-STYLE QUESTIONS

Question 7

Select the strongest type of bond.

A. ionic

B. metallic

C. covalent

D. dispersion forces

Question 8

Which of the following ionic compounds is insoluble in water?

A. potassium sulfate

B. iron nitrate

C. ammonium hydroxide

D. magnesium carbonate

Question 9

When silver nitrate reacts with sodium chloride, a precipitate forms according to the equation:

$AgNO_3(aq) + NaCl(aq) \rightarrow AgCl(s) + NaNO_3(aq)$

Identify the spectator ions.

A. Ag^+ and Cl^-

B. Na^+ and Cl^-

C. Na^+ and NO_3^-

D. Ag^+ and NO_3^-

Question 10

Under a particular set of conditions of thin-layer chromatography, component A adsorbs more strongly to the polar stationary phase than component B. Which of the following options best describes their polarity and retention factor?

A. Component A is more polar than component B and will have a higher retention factor.

B. Component A is less polar than component B and will have a higher retention factor.

C. Component A is more polar than component B and will have a lower retention factor.

D. Component A is less polar than component B and will have a lower retention factor.

Short-answer questions

Question 1

The electronegativities of C, H and Cl are 2.6, 2.1 and 3.2 respectively. What is the most polar bond out of C–Cl, C–H and H–Cl? Justify your answer. 3 marks

EXAM-STYLE QUESTIONS

Question 2

Briefly explain each of the following statements.

a. A chlorine atom has a smaller atomic radius than a sodium atom. 2 marks

b. Graphite and diamond are both composed only of carbon atoms. Graphite can conduct electricity, but diamond cannot. 2 marks

c. Ionic compounds cannot conduct electricity in the solid state but can when molten. 2 marks

d. All bonding in covalent molecules and ionic compounds and metals is due to electrostatic attraction. 3 marks

Question 3

Consider the compounds carbon dioxide, CO_2, and hydrogen sulfide, H_2S. Predict whether these compounds are polar or non-polar. Justify your prediction in terms of the nature of bonding within each molecule and the shape of each molecule. Illustrate your answer by using appropriate diagrams. 6 marks

ISBN 978 0 6557 0015 9

EXAM-STYLE QUESTIONS

Question 4

A student sets up a paper chromatogram and places a spot of green food dye on the origin. In one experiment, the solvent moved 12 cm and a blue spot moved 9 cm from the origin. A yellow spot moved 5 cm. When the experiment was repeated, the blue spot moved 15 cm.

a. Determine how far (in cm) from the origin the solvent is likely to be in the second experiment. 2 marks

b. Paper is made from polar molecules. Identify whether the blue or yellow food dye is more polar. Explain your answer. 2 marks

Question 5

a. Complete the table by writing the formula of each of the ionic compounds listed. 3 marks

Name of compound	Formula of compound
barium iodide	
sodium nitride	
calcium hydroxide	

b. Complete the table by writing the name of each of the ionic compounds listed. 3 marks

Formula of compound	Name of compound
K_2O	
Na_2SO_4	
CuCl	

c. Phosphate is a polyatomic ion that contains phosphorus and oxygen.

i. Write the formula of the phosphate ion. 1 mark

ii. Phosphorus is referred to as a critical element. Explain the meaning of this term and why phosphorus is classified this way. 2 marks

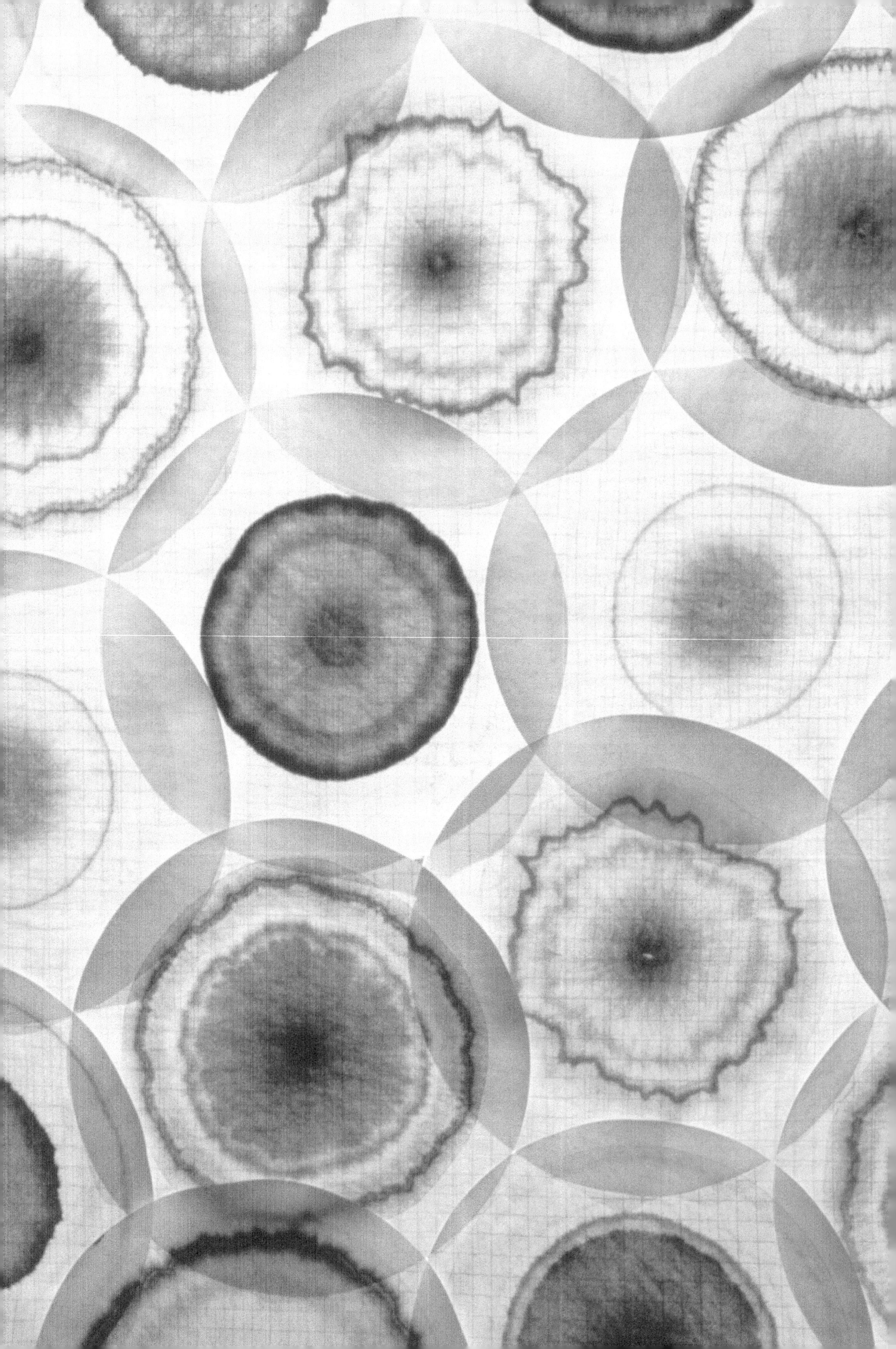

UNIT 1

How can the diversity of materials be explained?

AREA OF STUDY 2

How are materials quantified and classified?

Outcome

On completion of this unit the student should be able to calculate mole quantities, use systematic nomenclature to name organic compounds, explain how polymers can be designed for a purpose, and evaluate the consequences for human health and the environment of the production of organic materials and polymers.

Key knowledge

Quantifying atoms and compounds

- the relative isotopic masses of isotopes of elements and their values on the scale in which the relative isotopic mass of the carbon-12 isotope is assigned a value of 12 exactly
- determination of the relative atomic mass of an element using mass spectrometry (details of instrument not required)
- Avogadro's constant as the number 6.02×10^{23} indicating the number of atoms or molecules in a mole of any substance; determination of the amount, in moles, of atoms (or molecules) in a pure sample of known mass
- determination of the molar mass of compounds, the percentage composition by mass of covalent compounds, and the empirical and molecular formula of a compound from its percentage composition by mass

Families of organic compounds

- the grouping of hydrocarbon compounds into families (alkanes, haloalkanes, alkenes, alcohols, carboxylic acids) based upon similarities in their physical and chemical properties, including general formulas and general uses based on their properties
- representations of organic compounds (structural formulas, semi-structural formulas) and naming according to the International Union of Pure and Applied Chemistry (IUPAC) systematic nomenclature (limited to non-cyclic compounds up to C8, and structural isomers up to C5)
- plant-based biomass as an alternative renewable source of organic chemicals (for example, solvents, pharmaceuticals, adhesives, dyes and paints) traditionally derived from fossil fuels
- materials and products used in everyday life that are made from organic compounds (for example, synthetic fabrics, foods, natural medicines, pesticides, cosmetics, organic solvents, car parts, artificial hearts), the benefits of those products for society, and the health and/or environmental hazards they pose

Polymers and society

- the differences between addition and condensation reactions as processes for producing natural and manufactured polymers from monomers
- the formation of addition polymers by the polymerisation of alkene monomers
- the distinction between linear (thermoplastic) and cross-linked (thermosetting) addition polymers with reference to structure and properties
- the features of linear addition polymers designed for a particular purpose, including the selection of a suitable monomer (structure and properties), chain length and degree of branching

- the categorisation of different plastics as fossil fuel-based (HDPE, PVC, LDPE, PP, PS) and as bioplastics (PLA, Bio-PE, Bio-PP); plastic recycling (mechanical, chemical, organic), compostability, circularity and renewability of raw ingredients
- innovations in polymer manufacture using condensation reactions, and the breakdown of polymers using hydrolysis reactions, contributing to the transition from a linear economy towards a circular economy

Quantifying atoms and compounds

Atoms are so small that they cannot readily be counted individually or even in the thousands of millions. Instead, chemists use special units of measurement to record quantities of atoms, elements and compounds.

RELATIVE MASS

The masses of single atoms and molecules are incredibly small, not easily measured and can be inconvenient to use in calculations. Hence, the masses used in chemistry are relative masses. They are relative to the standard of the common isotope carbon-12 (^{12}C), which is assigned a mass of exactly 12.

- **Relative isotopic mass** is the relative mass of an individual isotope of an element.
- **Relative atomic mass (A_r)** is the weighted average of the relative masses of the naturally occurring isotopes of a particular element. The relative atomic mass of each element is included in the periodic table.

Table 1.2.1 shows the relative masses of the two naturally occurring copper isotopes. as well as their relative atomic mass.

Table 1.2.1 Naturally occurring isotopes of copper

Element	Isotope	Relative isotopic mass	Relative atomic mass
copper	^{63}Cu	62.9	63.5 (as appears on the periodic table)
	^{65}Cu	64.9	

Mass spectrometry

Relative atomic mass can be determined using a mass spectrometer to identify the existence of isotopes of an element and their relative isotopic abundances.

The data from a mass spectrometer is in the form of a **mass spectrum** (Figure 1.2.1).

Each peak on the spectrum represents one isotope. The position of each peak indicates the relative isotopic mass, and the height of the peak indicates its relative isotopic abundance.

Looking at the mass spectrum data in Figure 1.2.1:

- one isotope has a relative mass of 62.9 and relative abundance of 69.1%
- one isotope has a relative mass of 64.9 and relative abundance of 30.9%.

The average relative atomic mass of a copper atom is then calculated:

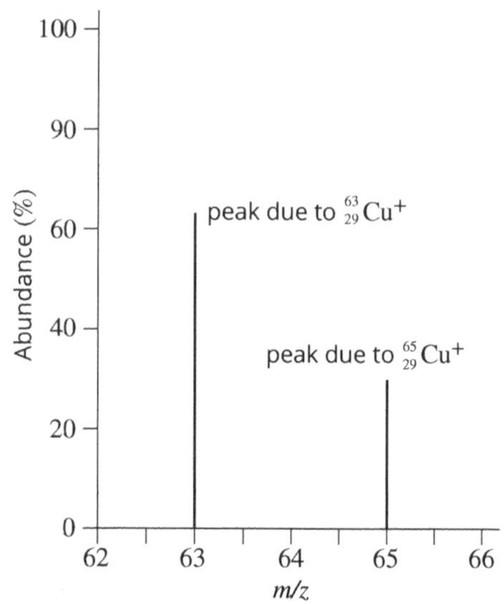

Figure 1.2.1 The mass spectrum of copper

$$A_r(\text{Cu}) = \frac{(\%\text{ abundance} \times \text{relative isotopic mass}) + (\%\text{ abundance} \times \text{relative isotopic mass})}{100}$$

$$= \frac{(69.1 \times 62.9) + (30.9 \times 64.9)}{100}$$

$$= 63.5$$

 ISBN 978 0 6557 0015 9

KEY KNOWLEDGE

AVOGADRO'S CONSTANT AND THE MOLE

Chemists use a special unit of measurement called the mole. One mole of any substance contains Avogadro's constant of particles. **Avogadro's constant** is the number 6.02×10^{23}. Table 1.2.2 provides examples of how the mole is used as a counting unit.

Table 1.2.2 Examples of the use of the mole

Number of moles	Number and type of particles
1.00 mol of Na atoms	6.02×10^{23} Na atoms
1.00 mol of H_2O molecules	6.02×10^{23} H_2O molecules 6.02×10^{23} O atoms because there is one O atom per molecule $2 \times 6.02 \times 10^{23} = 1.20 \times 10^{24}$ H atoms because there are two H atoms per water molecule
3.00 mol of O_2 molecules	$3 \times 6.02 \times 10^{23} = 1.81 \times 10^{24}$ O_2 molecules $2 \times 1.81 \times 10^{24} = 3.61 \times 10^{24}$ O atoms

It is very difficult to conceptualise just how big a number Avogadro's constant is, especially when atoms and molecules are so small (Figure 1.2.2).

Figure 1.2.2 One mole of soft-drink cans would cover the Earth's surface to a depth of more than 320 km.

The mole is given the symbol n and the unit mol. A useful formula links the amount of substance (n), the number of particles in the substance (N) and Avogadro's number (N_A) of 6.02×10^{23}:

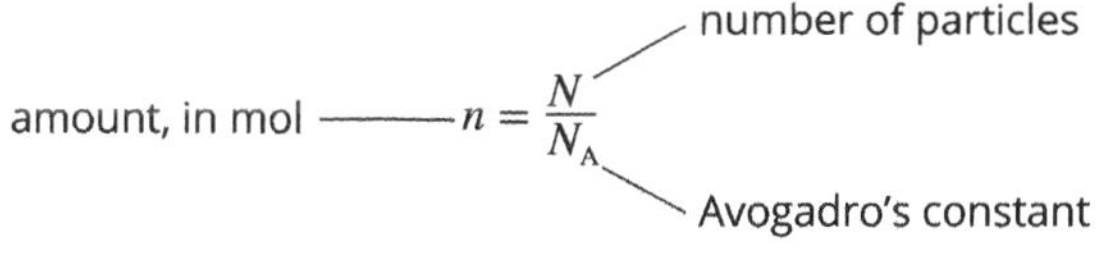

$$n = \frac{N}{N_A}$$

Molar mass

Molar mass is the mass of one mole of an element or compound. It is equal to the sum of the relative atomic masses of the atoms in the formula of the substance expressed in grams per mole. Molar mass is given the symbol M and the unit $g\,mol^{-1}$. Some examples are provided in Table 1.2.3.

Table 1.2.3 Calculation of molar masses of some common elements and compounds

Substance	Relative atomic masses	Molar mass of substance
Na	Na = 23.0	$23.0\,g\,mol^{-1}$
O_2	O = 16.0	$2 \times 16.0 = 32.0\,g\,mol^{-1}$
$NaNO_3$	Na = 23.0 N = 14.0 O = 16.0	$23.0 + 14.0 + (3 \times 16.0) = 85.0\,g\,mol^{-1}$

Molar mass is used to determine the amount in moles of substance in a pure sample of known mass. A useful formula links the amount of a substance (n), its molar mass (M) and the given mass of a substance (m):

amount, in mol — $n = \frac{m}{M}$ — mass, in g; molar mass, in g mol^{-1}

- **You will now be able to complete Worksheets 13–15 and conduct Practical activity 8.**

FORMULAS OF COMPOUNDS

The relative proportions of each element in a covalent compound can be expressed as a:

- percentage in terms of the mass contributed by each element
- empirical formula in terms of the simplest ratio of atoms contributed by each element
- molecular formula in terms of the actual number of atoms of each element present in the compound.

Percentage composition

The **percentage composition** by mass of elements in a compound of known formula can be determined by using relative atomic masses:

$$\%\text{ by mass} = \frac{\text{mass of the element in 1 mol of compound}}{\text{molar mass of the compound}} \times 100$$

Example: Find the percentage composition by mass of each element in the common compound NH_4NO_3.

$$\begin{aligned}\%N &= \frac{2\times14.0}{2\times14.0+4\times1.0+3\times16.0}\times100\\ &= \frac{28.0}{80.0}\times100\\ &= 35.0\%\end{aligned}$$

$$\%H = \frac{4\times1.0}{80.0}\times100$$
$$= 5.0\%$$

$$\%O = \frac{3\times16.0}{80.0}\times100$$
$$= 60.0\%$$

Empirical formulas

Atoms or ions exist in compounds in fixed whole-number ratios. The **empirical formula** of a compound is the simplest whole-number ratio of elements in that compound.

The empirical formula of any compound is determined using the mass of each element present in a given mass of the compound. These masses can be determined experimentally.

Example: **Determine the empirical formula of a compound, a sample of which has been experimentally determined to contain 82.75% carbon and 17.25% hydrogen.**

Thinking	Working	
1 Determine the mass of each element present in 100 g of substance from the percentage compositions and write the ratio by mass.	C : H 82.75 g : 17.25 g	
2 Calculate the amount, in mol, of each element present using $n = \frac{m}{M}$ to determine a mole ratio.	$n(C) = \frac{82.75}{12.0}$ $= 6.896$	$n(H) = \frac{17.25}{1.0}$ $= 17.25$
3 Divide each by the smaller amount to simplify the ratio.	$= \frac{6.896}{6.896}$ $= 1$	$= \frac{17.25}{6.896}$ $= 2.5$
4 Find a multiple to give a whole-number ratio.	$= 1 \times 2$ 2 : 5	$= 2.5 \times 2$
5 Write the empirical formula.	C_2H_5	

Molecular formulas

The **molecular formula** of a compound gives the actual number of atoms present in a molecule. It can be the same as, or different from, the empirical formula (Table 1.2.4).

Molecular formulas are calculated from the empirical formula and molar mass of a compound. The molecular formula is always a whole-number multiple of the empirical formula.

Table 1.2.4 Some molecular and empirical formulas of organic compounds

Molecule	Molecular formula	Empirical formula
ethane	C_2H_6	CH_3
propanol	C_3H_8O	C_3H_8O
glucose	$C_6H_{12}O_6$	CH_2O

Example: **Determine the molecular formula of a compound with the empirical formula C_2H_5 and a molecular mass of 58.0 g mol^{-1}.**

Thinking	Working
1 Find the molar mass of the empirical formula (EF).	$M(EF) = (2\times12.0)+(5\times1.0)$ $= 29.0\,g\,mol^{-1}$
2 Determine the number of empirical formula units in the molecular formula by dividing the molar mass of the molecular formula by the molar mass of the empirical formula.	Number of EF units $= \frac{M(EF\ unit)}{M(molecule)}$ $= \frac{58.0}{29.0}$ $= 2$
3 Write the molecular formula.	$= 2 \times C_2H_5$ $= C_4H_{10}$

- **You will now be able to complete Worksheet 16 and conduct Practical activities 9 and 10.**

 ISBN 978 0 6557 0015 9

Families of organic compounds

Organic compounds are compounds whose molecules contain carbon covalently bonded to atoms of other elements.

HYDROCARBON FAMILIES

Hydrocarbons are organic compounds that contain only carbon and hydrogen. Hydrocarbon molecules are held together by covalent bonds. A group of hydrocarbons in which successive members differ by $-CH_2-$ is considered a family and called a **homologous series**. Members of a hydrocarbon family have similarities in structure, which give them similarities in their physical and chemical properties. Alkanes and alkenes are two of the simplest families of hydrocarbons (Table 1.2.5). **Alkanes** are described as saturated because the contain all single covalent bonds. **Alkenes** contain a double carbon–carbon bond and are described as unsaturated.

Representing hydrocarbons

There are different ways of representing hydrocarbons and other organic molecules. **Structural formulas** show all bonds present in the molecule, whereas **semi-structural formulas** show the atoms bonded to each carbon atom written linearly (Table 1.2.6).

Table 1.2.5 Alkanes and alkenes

Hydrocarbon family	General formula	Characteristics	Example of structural formulas
Alkanes • contain all single covalent bonds	C_nH_{2n+2}	• saturated • non-polar • weak dispersion forces between molecules that increase in strength as molecules get bigger	methane, ethane, propane
Alkenes • contain a carbon–carbon double bond	C_nH_{2n}	• unsaturated • non-polar • weak dispersion forces between molecules that increase in strength as molecules get bigger	ethene, propene

Table 1.2.6 Representations of organic molecules

Molecular formula	Type of molecule	Structural formula	Semi-structural formula
C_3H_8	alkane		$CH_3CH_2CH_3$
C_3H_6	alkene		CH_2CHCH_3

KEY KNOWLEDGE

Structural isomers

Structural isomers are compounds with the same composition and relative molecular mass, hence molecular formula, that differ in their structural and semi-structural formulas (Figure 1.2.3).

Figure 1.2.3 The structural isomers of C_4H_{10}

HALOALKANES, ALCOHOLS AND CARBOXYLIC ACIDS

Haloalkanes, alcohols and carboxylic acids are also families of organic molecules. Each family contains molecules with a characteristic **functional group**.

Functional groups

Functional groups are atoms or groups of atoms attached to a molecule, which determine the properties of the molecule (Table 1.2.7).

Table 1.2.7 Names and structures of some common functional groups

Name of functional group	Structure
halogen	–Cl, –F, –Br, –I
hydroxyl	–OH
carboxyl	–COOH

In **haloalkanes**, the functional group is a halogen atom. Halogen atoms are in group 17 of the periodic table and include chlorine, fluorine, iodine and bromine. The term **alcohol** is used for any hydrocarbon containing a hydroxyl group. **Carboxylic acids** contain the carboxyl functional group (Table 1.2.8).

Table 1.2.8 Homologous series of chloroalkanes, alcohols and carboxylic acids

Series	Molecule	Semi-structural formula
chloroalkanes (a type of haloalkane)	chloromethane	CH_3Cl
	chloroethane	CH_3CH_2Cl
	1-chloropropane	$CH_3CH_2CH_2Cl$
	1-chlorobutane	$CH_3CH_2CH_2CH_2Cl$
alcohols	methanol	CH_3OH
	ethanol	CH_3CH_2OH
	propan-1-ol	$CH_3CH_2CH_2OH$
	butan-1-ol	$CH_3CH_2CH_2CH_2OH$
carboxylic acids	methanoic acid	HCOOH
	ethanoic acid	CH_3COOH
	propanoic acid	CH_3CH_2COOH
	butanoic acid	$CH_3CH_2CH_2COOH$
Note: Only the first four molecules in each series are listed but you are expected to learn the names to C_8 and structural isomers up to C_5.		

IUPAC NAMING OF ORGANIC COMPOUNDS

In order to name organic compounds unambiguously, an international set of rules has been devised called the International Union of Pure and Applied Chemistry (IUPAC) systematic nomenclature.

- The first part of the name refers to the number of carbon atoms in the longest chain in the molecule (Table 1.2.9).

Table 1.2.9 Terms used in systematic naming of straight-chain hydrocarbons

No. carbon atoms	Term
1	meth-
2	eth-
3	prop-
4	but-
5	pent-
6	hex-
7	hept-
8	oct-

- The second part of the name refers to the homologous series of which the molecule is a member.

 ISBN 978 0 6557 0015 9

KEY KNOWLEDGE

- If a structural isomer is possible, the location of the double bond or functional group is indicated by giving the number of the first carbon atom involved in the double bond (Figure 1.2.4a), or the carbon to which the functional group is attached (Figure 1.2.4b).

(a)

(b)

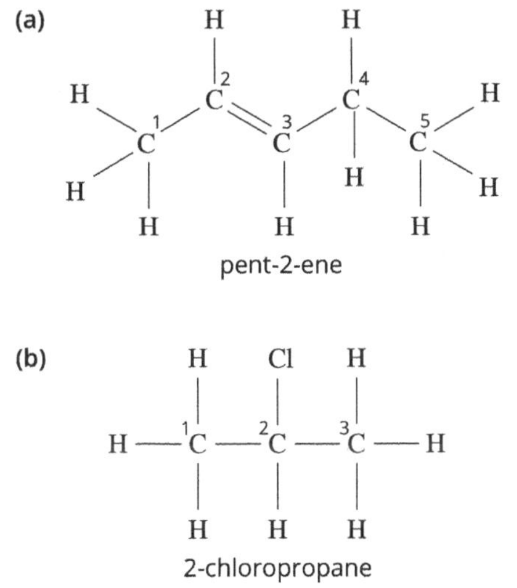

Figure 1.2.4 The structural formulas for (a) pent-2-ene and (b) 2-chloropropane

- The location of side branches, called alkyl groups, is indicated by adding the name of the group and the number of the carbon to which the group is attached before the molecule's name (Figure 1.2.5).
  - the first part of the name of the group is the number of carbons in the side branch
  - the last part of the name is *-yl*.

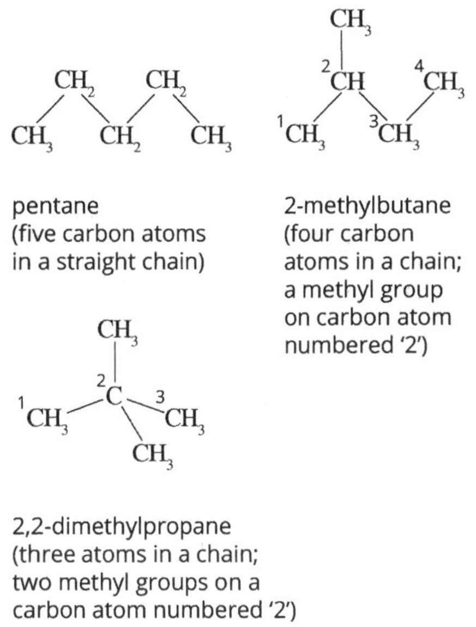

Figure 1.2.5 The names and isomers of C_5H_{12}

- If a structural isomer of a haloalkane is possible, the location of the functional group is indicated by including the number of the carbon to which the functional group is attached (Figure 1.2.4b).

A summary of IUPAC rules for naming compounds is in Table 1.2.10.

Table 1.2.10 IUPAC rules for naming organic compounds

Rule	Example	Name
1 Choose the longest carbon chain and name according to the alkane with the same number of carbon atoms.	$CH_3CH_2CH_2CH_2CH_2CH_3$	hexane
2 If there is a carbon–carbon double bond, replace *-ane* with -number from the end that gives the smallest number for the first carbon atom involved in the bond.	$CH_3CH{=}CHCH_2CH_2CH_3$	hex-2-ene (not hex-4-ene)
3 If there is a branch or functional group, number from the end that gives the smallest number.	$CH_3CH_2CH(CH_3)CH_3$	2-methylbutane (not 3-methylbutane)
4 Number all branches. Use prefixes such as *di-*, *tri-* and *tetra-* if branches are identical.	$CH_3CH(CH_3)CH(CH_3)CH_3$	2,3-dimethylbutane
5 If different alkyl branches are present, write them in alphabetical order.	$CH_3C(C_2H_5)(CH_3)CH_2CH_2CH_3$	2-ethyl-2-methylpentane
6 For haloalkanes, add the name of the atom as *halo-* as a prefix to the molecule and if isomers exist, indicate the location of the halogen atom with a number.	$CH_3CH_2CH_2Cl$	1-chloropropane
7 For alcohols, replace -e with *-ol* and, if isomers exist, indicate the location of the hydroxy group with a number.	$CH_3CH_2CH_2OH$	propan-1-ol
8 For carboxylic acids, replace -e with *-oic acid*.	CH_3CH_2COOH	propanoic acid

KEY KNOWLEDGE

Table 1.2.11 Physical properties of alkanes, alkene, haloalkanes, alcohols and carboxylic acids

Homologous series	Polarity	Strongest Intermolecular force	Melting and boiling points	Solubility in water	Solubility in non-polar solvents
alkanes and alkenes	non-polar	dispersion	• relatively low • increase as mass of molecule increases	insoluble	highly soluble
haloalkanes	halogen group is polar, hydrocarbon ends are non-polar	dipole–dipole attractions	• higher than for alkanes and alkenes	Small molecules are slightly soluble in water. Solubility decreases as length of hydrocarbon tail increases.	Solubility increases as length of hydrocarbon tail increases.
alcohols and carboxylic acids	–OH and –COOH groups are polar, hydrocarbon ends are non-polar	hydrogen bonds	• higher than for alkanes and alkenes • carboxylic acids have highest melting and boiling points	Small molecules are highly soluble. Solubility decreases as length of hydrocarbon tail increases.	Solubility increases as length of hydrocarbon tail increases.

PHYSICAL AND CHEMICAL PROPERTIES

The melting and boiling points of organic molecules are determined by the size and type of intermolecular forces (Table 1.2.11).

The solubility of organic molecules in polar water and non-polar organic solvents is determined by the molecule's ability to form hydrogen bonds, dipole–dipole forces or only dispersion forces with the molecules of the solvent.

The chemical properties of these organic compounds include:

- The carbon–carbon double bond in alkenes allows alkenes to undergo addition reactions, in which the double bond breaks to become a single bond and an atom or group of atoms is added to both carbon atoms involved in the double bond. There are no by-products. When an alkene reacts with bromine, as in Figure 1.2.6, it causes the bromine to discolour, making this a useful test for the presence of alkenes.

$$CH_2=CH_2 + Br_2 \longrightarrow CH_2Br{-}CH_2CH_2Br$$

1,2-dibromoethane

Figure 1.2.6 The presence of a carbon–carbon double bond can be indicated by reacting the substance with bromine. Only an alkene will discolour the bromine.

SOURCES AND USES OF ORGANIC COMPOUNDS

Many products used by society in a huge variety of ways contain organic compounds. These include solvents, pharmaceuticals, adhesives, dyes and paints.

Fossil fuels

Fossil fuels such as crude oil, coal and natural gas are naturally occurring mixtures of hydrocarbons that have accumulated from the remains of dead organisms over millions of years in the Earth's crust. Fossil fuels are:

- non-renewable because they are produced over a very long time period
- currently the world's most important energy sources
- currently the world's most important sources of naturally occurring alkanes and alkenes as well as raw materials used in the production of haloalkanes, alcohols and carboxylic acids.

Plant-based biomass

Plant-based **biomass** is a carbon-based material that comes from plants. The substances that make up biomass are renewable because they can be replenished quickly from natural sources. Many compounds derived from plants contain additional elements to carbon and hydrogen so provide different compounds as raw materials that may be more useful than the hydrocarbons in fossil fuels.

Common organic molecules that are being produced from raw materials from plants include ethanol and ethene.

 ISBN 978 0 6557 0015 9

Table 1.2.12 Materials and products used in everyday life that are made from organic compounds

Homologous series	Example of products	Benefits to society	Health or environmental hazards
alkanes	propane in gas bottles	a cheap and simple production of heat which can be used for cooking or heating	Combustion of propane releases carbon dioxide to the atmosphere, which contributes to the enhanced greenhouse effect.
alkenes	polymers including polyethene	a huge range of applications including food storage, synthetic fabrics, car parts, artificial hearts	Only some polymers can be recycled.
haloalkanes	chlorofluorocarbons (CFCs)	• coolants in refrigeration • propellants in aerosols • foaming agents in fire extinguishers • solvents	CFCs contribute to the destruction of the ozone layer in the atmosphere, which allows increased levels of UV light to reach the Earth. This is associated with increased incidence of skin cancer and cataracts.
alcohols	ethanol	alcoholic drinks, including beer, wine and champagne	High alcohol consumption is associated with detrimental health effects.
carboxylic acids	vinegar	common ingredients in many foods and used as a natural preservative and disinfectant	There are no hazards of significance.

Uses, benefits and potential impacts

Our society is heavily reliant on organic compounds for products used in diverse applications such as foods, medicines, pesticides, cosmetics, solvents, fabrics and household implements. Shifting the primary source of organic compounds that are raw materials for these products from non-renewable fossil fuels to renewable plant-based biomass will help reduce the environmental impacts. A specific use of each type of organic compound is outlined in Table 1.2.12.

- **You will now be able to complete Worksheets 17 and 18 and conduct Practical activities 11 and 12.**

Polymers and society

Polymers are very large covalent molecular substances that:

- consist of monomers linked together
- contain thousands of atoms
- contain molecules of different sizes, consisting of different numbers of atoms
- are produced by a process called polymerisation
- exist naturally in living things or are manufactured for commercial use.

Many polymers are commonly known as plastics in society.

There are thousands of different polymers with different properties. However, in general, polymers are:

- lightweight
- non-conductors of electricity
- acid resistant
- flammable.

PRODUCTION OF POLYMERS

Polymers are produced by linking together small molecules called **monomers** in a process called **polymerisation**. The linking occurs through addition or condensation reactions to produce different types of polymers (Table 1.2.13).

Manufactured polymers are produced by both addition and condensation polymerisation for different purposes.

Natural polymers such as carbohydrates and proteins are produced by condensation polymerisation.

Table 1.2.13 Comparison of addition and condensation polymerisation

	Monomer	Features of process	Feature of polymer
addition polymerisation	unsaturated alkene e.g. ethene 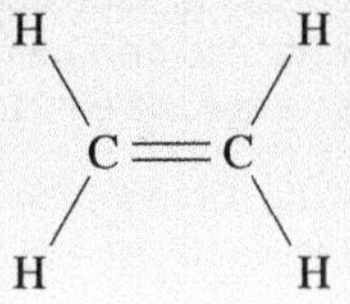	No other products are formed.	Polymer chain contains only carbon atoms and only single carbon–carbon bonds.
diagram	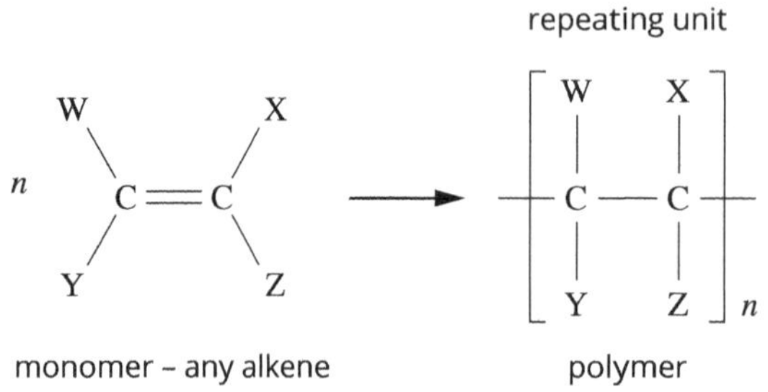 repeating unit monomer – any alkene polymer		
	Monomer	**Features of process**	**Feature of polymer**
condensation polymerisation	functional group on each end of monomer, e.g. 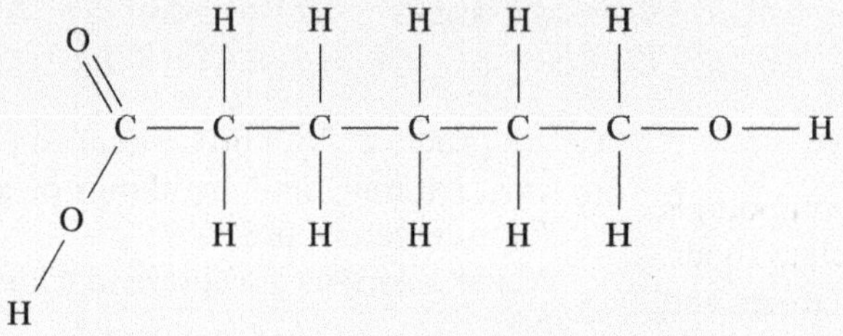	Functional groups react. Small molecules such as water are also formed.	Polymer chain can contain atoms other than carbon.
diagram	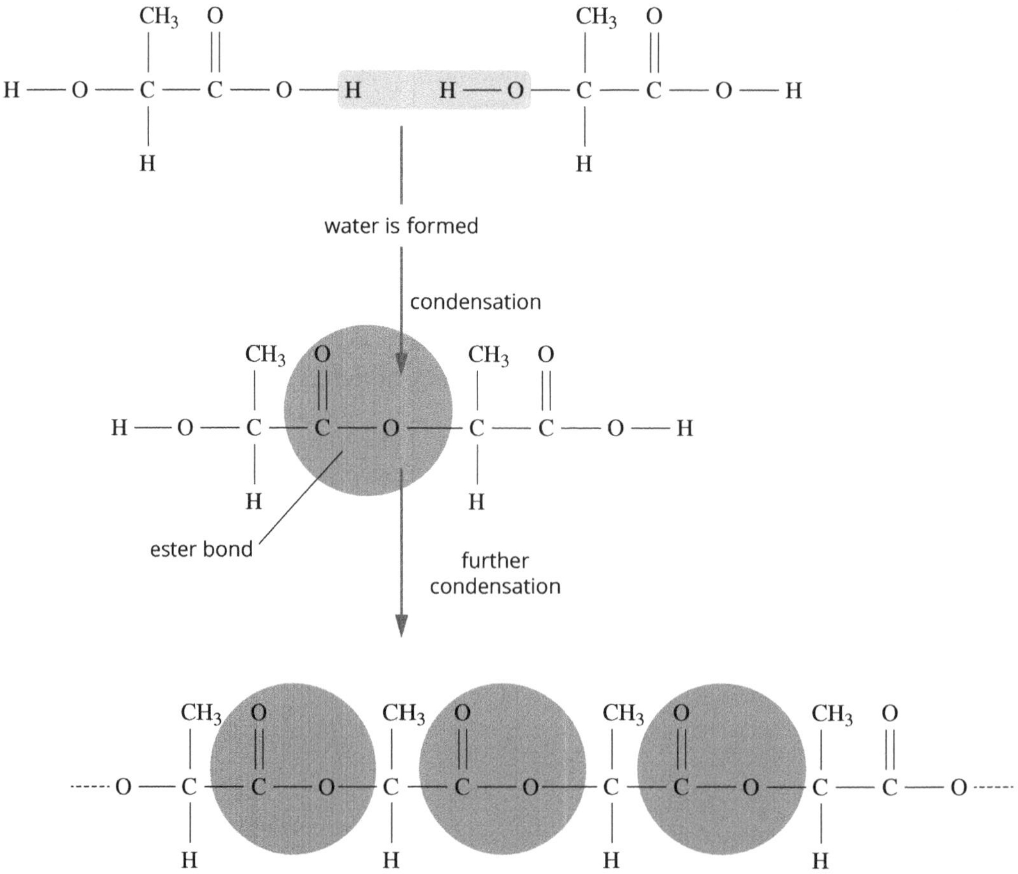 water is formed condensation ester bond further condensation		

 ISBN 978 0 6557 0015 9

KEY KNOWLEDGE

TYPES OF ADDITION POLYMERS

Different types of polymers can be produced with different properties, making them suitable for various purposes. The differences include the:

- presence of cross-links between chains
- degree of branching in polymer chains
- type of monomer.

These differences have implications for whether a polymer can be recycled or composted.

Linear and cross-linked polymers

Linear (thermoplastic) and cross-linked (thermosetting) polymers are addition polymers that have different structures and properties according to the presence of cross-links (Table 1.2.14). Cross-links are covalent bonds that form between two different polymer chains. Cross-links are usually achieved by:

- heating the material after the polymer chains have been formed
- adding another substance that reacts with the atoms on the chains, joining the chains together.

Degree of branching

The amount of branching in addition polymers can also be varied by altering the production conditions. The properties of a polymer are affected by the degree of branching because this affects how closely the long polymer molecules can pack together. The more branching present, the less the chains can pack together and the lower the density. The degree of branching also effects properties such as melting point, strength and flexibility (Table 1.2.15).

Type of monomer

Monomers suitable for manufacturing of polymers are sourced from fossil fuels such as crude oil and natural gas, as well as plant-based biomass. Plastics produced from plants are known as bioplastics. Fossil fuels are non-renewable, hence they are a finite resource, whereas plants are renewable because they can be grown relatively quickly. Using different monomers produces polymers with different properties that are suitable for different purposes (Table 1.2.16).

Table 1.2.14 Linear and cross-linked polymers

Type of polymer	Structure	Properties	Applications
• linear • thermoplastic	• strong covalent bonds within chains • weak forces between different chains	• flexible • softens on heating so able to be remoulded	• items produced by moulding, e.g. laundry baskets, milk bottles, toys
• cross-linked • thermosetting	• strong covalent bonds within chains • strong covalent bonds called cross-links between different chains	• does not melt but chars when heated so cannot be remoulded • rigid	• laminates, saucepan handles

Table 1.2.15 Types of polyethene based on amount of branching present

Name	Branching	Production pressure	Properties
high-density polyethene (HDPE)	very small amount	atmospheric pressure	stronger, less flexible, higher melting point
how-density polyethene (LDPE)	highly branched	high pressure	slightly weaker, more flexible, lower melting point

Table 1.2.16 Polymers manufactured and selected for different purposes

Name	Source of monomer	Monomer	Properties	Applications
HDPE (high-density polyethene)	fossil fuel	ethene $H_2C{=}CH_2$	stronger, less flexible	shampoo bottles, toys, milk jugs, recycling bins
LDPE (low-density polyethene)	fossil fuel	ethene	slightly weaker, more flexible	plastic bags, packaging, squeeze bottles
bio-PE (biopolyethene)	plant-based biomass	ethene (sourced from biomass)		
PVC (polyvinyl chloride)	fossil fuel	$HClC{=}CH_2$	higher melting point, low flammability, low electrical conductivity	water pipes, covering of electrical wires, cordial bottles
PP (polypropene)	fossil fuel	propene $H_2C{=}CH{-}CH_3$	durable, cheap	artificial grass, dishwasher-safe plastic, ice-cream containers, rope
bio-PP (biopolypropene)	plant-based biomass	propene (sourced from biomass)		
PS (polystyrene)	fossil fuel	styrene $C_6H_5CH{=}CH_2$	hard, brittle, low melting point	toys, packaging, expanded foams
PLA (polylactic acid)	plant-based biomass	lactic acid $CH_3{-}CH(OH){-}COOH$	dissolves in water	medical and cleaning uses

 ISBN 978 0 6557 0015 9

KEY KNOWLEDGE

DEALING WITH POLYMER WASTE

Although polymers offer many advantages to society, there are some significant environmental disadvantages to polymers including:

- most are currently made from monomers sourced from fossil fuels
- they take a long time to decompose
- when disposed of, they take up space in landfill
- they cause pollution in oceans, especially through formation of micro plastics, which are often invisible, very small pieces of plastics
- produce toxic gases when combusted.

Plastic disposal and recycling

If a polymer is produced from raw materials sourced from a fossil fuel such as crude oil or natural gas, given a particular purpose, then disposed of for land fill, it is part of a linear economy. Polymers in landfill will take hundreds of years to decompose and can release toxic gases as they do so, which is an environmental hazard. Government and industry are seeking ways to improve recycling or compostability of polymers so that there is a shift to a circular economy and its associated sustainability benefits.

Only some plastics are compostable. Plant-based bioplastics are more compostable than those sourced from fossil fuels, meaning they are capable of disintegrating into natural elements in a compost without producing toxic products.

Different methods of recycling are used for different types of polymers (Table 1.2.17).

Table 1.2.17 Methods of recycling plastics

Type of recycling	Description
mechanical recycling	Mixed plastics are shredded to pellets, then remoulded to form new products such as garden furniture.
chemical recycling	Conditions such as catalysts and high temperature are used to change the structure of a polymer to form an oil that can be re-used to make more plastics.
organic recycling	Bacteria are used to break down polymer structures.

- **You will now be able to complete Worksheets 19–22 and conduct Practical activities 13 and 14.**

Knowledge review—comparing metallic, ionic and covalent bonding models

1 Write each term in the table, next to its definition.

ionic bond	cation	covalent bond	lattice
molecule	non-bonding electrons	metallic bond	

	positively charged ion
	electrostatic attraction between a positive ion and a negative ion
	electrostatic attraction between one or more pairs of shared electrons and two positive nuclei
	regular 3D arrangement of large numbers of atoms or ions
	discrete particle containing two or more atoms covalently bonded
	valence electrons that are not involved in bonding
	electrostatic attraction between positive ions and negative delocalised electrons

2 A chemistry student is given a solid sample of an unknown material. The student carries out the following tests to determine the type of bonding present. Fill in the expected result for each test for the different models of bonding.

Test	Metallic structure	Ionic lattice	Covalent molecular structure	Covalent lattice structure
electrical conductivity of the solid samples				
electrical conductivity of the samples in molten state				
melting temperature				

 ISBN 978 0 6557 0015 9

WORKSHEET 14

Exploring relative mass

A chemistry student designed and carried out the following activity to gain an understanding of the concept of relative mass. The student used lollies to represent different elements of different mass.

The student:

- obtained samples containing 50 of each of the following lollies: fruit jubes, chocolate almonds, jelly beans and mint leaves
- used a balance to find the mass of each sample of 50 fruit jubes, chocolate almonds, jelly beans and mint leaves.

The student's records are shown in the second column of the table:

Sample	Mass of sample of 50 lollies (g)	Average mass of a single lolly (g)	Relative mass of a single lolly
fruit jubes	67.0		5.0
chocolate almonds	320.0		
jelly beans	88.0		
mint leaves	183.0		

1 Calculate the average mass (g) of a single lolly of each type and fill in the third column of the table.

2 A single fruit jube is taken as the standard and designated a mass of exactly 5.0. Comparison to this standard is then used to determine the relative mass of each of the other types of lollies using the following formula:

$$\text{relative mass of lolly} = \frac{\text{actual mass of lolly}}{\text{actual mass of single fruit jube}} \times 5.0$$

Calculate the relative masses of the other lollies and fill in the last column of the table.

3 Explain how this activity illustrates the difference between an actual mass and a relative mass.

4 What is the standard used in chemistry to determine all other relative masses?

5 In the student activity, the student used samples containing 50 lollies to find an average mass of a single lolly, rather than just measuring the mass of one lolly.

a Suggest an advantage to the student of taking this approach.

b Explain how this could relate to the concepts of relative isotopic and relative atomic mass.

WORKSHEET 15

Moles—the chemist's unit of measurement

1 The mole is a unit of measurement. It represents a physical quantity in much the same way as a more common unit of measurement, the dozen.

a Compare the use of the two units by calculating the actual numbers of the following amounts:

1 dozen = ______________   1 mol = ______________

4 dozen = 4 × 12 = ______________   4 mol = $4 \times 6.02 \times 10^{23}$ = ______________

0.10 dozen = ______________   0.10 mol = ______________

b Describe one similarity and one difference between the units 'mole' and 'dozen'.

2 Explain why a unit of measurement such as the mole is necessary in chemistry.

3 Examine the diagram, which shows a sample of common table sugar, sucrose ($C_{12}H_{22}O_{11}$), on a balance, and its mass in grams. The relative atomic masses of the elements found in sucrose are shown in the table.

Relative atomic masses of the elements found in sucrose	
Element	A_r
H	1.0
C	12.0
O	16.0

Complete the following table by first identifying the formula needed for each calculation and then solving for the answer. Don't forget to include the correct units in your answers. The first few formulas have been included for you.

Calculation	Relevant formula	Solution
molar mass of sucrose, $C_{12}H_{22}O_{11}$	$M(C_{12}H_{22}O_{11})$ = sum of relative atomic masses of elements in compound	
amount, in mol, of sucrose in the sample	$n(C_{12}H_{22}O_{11}) = \frac{m}{M}$	
number of sucrose particles in the sample	$N(C_{12}H_{22}O_{11}) = n \times N_A$	
amount, in mol, of carbon atoms in the sample	$n(C) = 12 \times n(C_{12}H_{22}O_{11})$	
number of carbon atoms in the sample		
total amount, in mol, of atoms in the sample		
total number of atoms in the sample		

 ISBN 978 0 6557 0015 9

WORKSHEET 16

Empirical and molecular formulas

1 The formulas of some chemical compounds are listed in the box. Circle the nine empirical formulas.

NaCl	CH_4	C_3H_6	$C_6H_{12}O_6$	C_5H_{10}
C_4H_{10}	$CaCl_2$	CO_2	CH_3COOH	H_2O
CH_3Cl	NH_3	NH_4NO_3	H_2SO_4	C_8H_{18}

2 In the table below, write the molecular formulas of the remaining six compounds, and then write the empirical formula of each in the second column.

Molecular formula	Empirical formula

3 Explain how the empirical formula of a substance can be the same as its molecular formula and give an example of a compound for which this is the case.

4 Compare the different amounts of carbon and hydrogen in the compounds listed in the table by calculating their respective percentages by mass.

Molecular formula	Empirical formula	Mass of carbon (%)	Mass of hydrogen (%)
CH_4			
C_3H_6			
C_5H_{10}			
C_8H_{18}			

5 Examine the percentage masses of the compounds that have the same empirical formula. What do you notice?

ISBN 978 0 6557 0015 9

WORKSHEET 17

Alkanes, alkenes and haloalkanes

The structures of 12 organic compounds are shown.

(A)
```
  H   H   H   H   H   H
  |   |   |   |   |   |
H-C - C = C - C - C - C-H
  |           |   |   |
  H           H   H   H
```

(B)
```
  H
  |
H-C-H
  |
  H
```

(C)
```
   Cl
   |
H--C--H
   |
   H
```

(D)
```
   I    H
   |    |
H--C----C--H
   |    |
   H    H
```

(E)
```
H     H
 \   /
  C=C
 /   \
H     H
```

(F)
```
   H    H    Br
   |    |    |
H--C----C----C--H
   |    |    |
   H    H    H
```

(G)
```
  H   H   H   H
  |   |   |   |
H-C - C - C - C-H
  |   |   |   |
  H   H   H   H
```

(H)
```
   H    Cl   H    H
   |    |    |    |
H--C----C----C----C--H
   |    |    |    |
   H    F    H    H
```

(I)
```
          H
          |
        H-C-H
  H       |   H    H    H
  |       |   |    |    |
H-C - C - C - C  - C-H
  |   |   |   |    |
  H   |   H   H    H
      |
    H-C-H
      |
      H
```

(J)
```
  H   H   H
  |   |   |
H-C - C - C-H
  |   |   |
  H   |   H
    H-C-H
      |
      H
```

(K)
```
  H   H   H   H   H   H
  |   |   |   |   |   |
H-C - C - C - C - C - C-H
  |   |   |   |   |   |
  H   H   H   H   H   H
```

(L)
```
  H   H   H   H   H   H   H   H
  |   |   |   |   |   |   |   |
H-C - C - C = C - C - C - C - C-H
  |   |           |   |   |   |
  H   H           H   H   H   H
```

1 List the letters of all the compounds that are unbranched alkanes.

2 List the letters of all the compounds that are branched alkanes.

3 List the letters of all the compounds that are alkenes.

4 List the letters of all the compounds that are haloalkanes.

5 a What is the meaning of the term 'homologous series'?

 ISBN 978 0 6557 0015 9

b Write the semi-structural formula of the next two members of a homologous series that come after CH_3CH_2Cl.

---

6 List the letters of all the compounds that are unsaturated.

---

7 List the letters of all the compounds that are polar.

---

8 List the letters of two compounds that are isomers.

---

9 Identify the compound you would expect to have the lowest boiling temperature and explain your answer.

---

10 Identify the compounds you would expect to decolourise bromine and explain your answer.

---

11 Give the IUPAC systematic name of each compound.

A ____

B ____

C ____

D ____

E ____

F ____

G ____

H ____

I ____

J ____

K ____

L ____

WORKSHEET 18

Families of organic molecules—haloalkanes, alcohols and carboxylic acids

Complete the following table. State whether each dot point statement is true or false by writing T or F in the appropriate column next to each dot point.

Name of molecule	Semi-structural formula	Structural formula	Statement	True or false
ethanol			is soluble in water	
			has the general formula C_nH_{2n}	
hexanoic acid			contains a C=C bond	
			contains a carboxyl group	
1,2-dichlorobutane			can be used as a solvent	
			is insoluble in water	
	$CH_3CH_2CH(OH)CH_3$		is an alcohol	
			does not form hydrogen bonds	
			is saturated	
	$CH_3CH_2CH_2COOH$		is soluble in water	
			is a member of the homologous series of alcohols	
			has only dispersion forces between molecules	
1-chloro-2-fluoropropane			contains two functional groups	
			is a polar molecule	
		Cl | Br—C—F | F	is highly flammable	
			contains two functional groups	
		H H | | O H—C—C—C// | | \\ H H O—H	contains a hydroxy functional group	
			forms hydrogen bonds	
		H | H—C—O—H | H	is formed by a reaction between an alcohol and a carboxylic acid	
			is a polar molecule	
			has a very high boiling point	

 ISBN 978 0 6557 0015 9

Polyethene—a case study of a polymer

1 Complete the following by naming and drawing the structural formula of the monomer of polyethene.

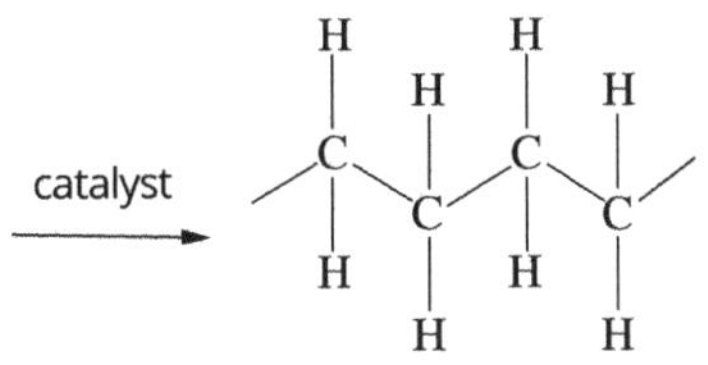

catalyst →

polythene

Monomer name:_______________

2 The monomer used to manufacture polyethene can be obtained from non-renewable and renewable sources.

a Identify one specific non-renewable and renewable source of the monomer and compare the environmental impacts of each.

Non-renewable source

____

Renewable source

____

Comparison of environmental impacts

____

____

____

b Does the source of the monomer make any difference to the compostability of polyethene? Explain your answer.

____

____

____

____

ISBN 978 0 6557 0015 9

3 Conditions during the manufacture of polyethene can be varied to make the product high density (HDPE) or low density (LDPE). Complete the table to compare the structure and properties of these two types of polyethene, and list four examples of each that you have in your home.

Type of polyethene	Structure	Properties	Four specific uses in your home
LDPE			
HDPE			

4 Identify and describe a method that can be used to recycle both types of polyethene.

5 Outline why many innovations in polymer manufacture are focused on transitioning from a linear to a circular economy.

 ISBN 978 0 6557 0015 9

WORKSHEET 20

Case study

Designing a polymer for a particular purpose

A chemical engineer wishes to manufacture a polymer for a client that is:

- high melting point
- somewhat flexible
- made from renewable raw materials
- able to be recycled and/or composted.

1 Use your knowledge of the features of different polymers to note what is required to produce the desired property.

Desired property	Implications for type and source of monomer, structure of polymer and end-of-life options
high melting point	
flexible	
sourced from raw materials	
able to be composted	

2 Suggest the name of a specific polymer that may be suitable for the chemist to manufacture. Evaluate how well this polymer would meet the client's request.

ISBN 978 0 6557 0015 9

Literacy review—naming compounds

1 Name the organic compounds in the table.

Molecular formula	Name
C_2H_6	
C_2H_4	
C_2H_5I	
C_2H_5OH	
CH_3COOH	

2 Write semi-structural formulas for the organic compounds in the table.

Compound	Semi-structural formula
propane	
propene	
1,2,3-trichloropropane	
propan-2-ol	
propanoic acid	

3 List the rules you used to name and write the formulas of the different organic compounds.

ISBN 978 0 6557 0015 9

WORKSHEET **22**

Reflection—How are materials quantified and classified?

The following table lists the key knowledge covered in this area of study.

1 Reflect on how well you understand the concepts listed. Rate your learning by shading the circle that corresponds to your current level of understanding for each one.

Key knowledge	Not confident ◄			► Very confident
Masses in chemistry—relative and actual, isotopic and atomic, use of mass spectrometry	○	○	○	○
The mole concept, Avogadro's constant and molar mass	○	○	○	○
Percentage composition, empirical formula and molecular formula	○	○	○	○
Families of organic compounds—alkanes, alkenes, haloalkanes, alcohols, carboxylic acids—structures, properties and uses	○	○	○	○
IUPAC systematic nomenclature of organic compounds	○	○	○	○
Addition and condensation polymers—different monomers and processes of polymerisation	○	○	○	○
Different types of plastics for different purposes	○	○	○	○
Recycling, composting and disposal of plastics to move from a linear to circular economy	○	○	○	○

2 Consider the points you have shaded from 'Not confident' to 'Very confident'. List specific ideas you can identify that were challenging.

3 Write down two different strategies that you will apply to help further your understanding of these ideas.

PRACTICAL ACTIVITY 8

Simulation

Mole simulations and applications

SUGGESTED DURATION

- 50 minutes

INTRODUCTION

The mole is a counting unit used in chemistry. The molar mass of an element or compound is the mass of one mole of the element or compound. Knowledge of molar masses allows chemists to weigh out a specific amount of a substance.

AIM

- To use mass as a means of counting the number of particles in a sample
- To apply the mole concept to some everyday scenarios

PRE-LAB SAFETY INFORMATION
Make sure you drink only the water provided specifically for the experiment in Part C.
Please indicate that you have understood the information in the safety table. Name (print): I understand the safety information (signature):

MATERIALS

Part A

- 2 small disposable paper cups
- jelly beans
- Nestle® Smarties
- a mixture of jelly beans prepared by your teacher
- a mixture of Smarties prepared by your teacher
- electronic balance (2 decimal places)

Part B

- disposable paper cup
- electronic balance (2 decimal places)

PART A • JELLY BEANS AND SMARTIES: A MOLE SIMULATION

METHOD

For this experiment, eight sweets will be referred to as a 'goop'.

1. Weigh accurately 1 goop of jelly beans in a clean paper cup. Record the mass.
2. Weigh accurately 1 goop of Smarties in another clean paper cup. Record the mass.
3. Your teacher will provide you with a sample of jelly beans and a separate sample of Smarties. Weigh the mixtures accurately. Record the masses in the Table 1.
4. Return used equipment.

RESULTS

Table 1 Masses of samples of Smarties and jelly beans

Sample	Mass (g)
1 goop of jelly beans	
1 goop of Smarties	
sample of jelly beans	
sample of Smarties	

ISBN 978 0 6557 0015 9

DISCUSSION

1 How many jelly beans are there in 1 goop of jelly beans?

2 How many Smarties are there in 3 goops of Smarties?

3 How many sweets are there in 2 goops of jelly beans and 1.5 goops of Smarties?

4 What is the relationship between the number of sweets and the amount of sweets in goops?

5 The mass of 1 goop of sweets is called its 'goopar' mass. What is the goopar mass of jelly beans?

6 What is the goopar mass of Smarties?

7 What would be the mass of 3 goops of jelly beans?

8 What is the mass of 2.5 goops of Smarties?

9 If you had 100 jelly beans, what amount of jelly beans would you have, in goops?

10 If you had 250 g of jelly beans, what mass of Smarties would contain the same number of sweets?

11 What is the relationship between the mass of jelly beans, the amount of goops, and the goopar mass of jelly beans?

12 a Without counting the jelly beans and Smarties in your samples, predict the numbers of jelly beans and Smarties.

b Count the jelly beans and Smarties and compare to your predictions.

13 In this model of moles, what do 'goops' and 'goopar' represent?

14 Is this a good model for the mole theory? Why or why not? Ensure to make note of the positive and negatives of the representation of mole theory in this model.

PART B • HOW MANY WATER MOLECULES ARE IN A MOUTHFUL OF WATER?

METHOD

1. Half-fill a new paper cup with water.
2. Weigh the cup and contents on an electronic balance and record the mass in Table 2.
3. Drink a single mouthful of water.
4. Weigh the cup with the remaining water. Record its mass.
5. Return used equipment.

RESULTS

Table 2 Masses of water

Mass of polystyrene cup and water	Mass (g)
before drinking a mouthful	
after drinking a mouthful	

DISCUSSION

1 What was the mass of your mouthful of water?

2 Calculate the amount, in moles, of water in your mouthful of water.

3 Calculate the number of water molecules in your mouthful of water.

4 Comment on the number of water molecules in your mouthful. What does this tell you about the size of water molecules?

CONCLUSION

 ISBN 978 0 6557 0015 9

Determining the molar mass of an element and a compound

SUGGESTED DURATION

- 20 minutes for Part A plus several hours standing time and drying time overnight
- 40 minutes for Part B plus drying time overnight

MATERIALS

- 2 g solid copper(II) oxide (black), CuO
- 3 g zinc granules or powder, Zn
- 50 mL of 2 M sulfuric acid, H_2SO_4
- 250 mL glass beaker
- weighing bottle/weight boat
- evaporating dish
- electronic balance (3 or 4 decimal places)
- Bunsen burner, tripod and gauze mat
- 50 mL glass measuring cylinder
- glass stirring rod
- spatula
- safety glasses/goggles
- disposable gloves

INTRODUCTION

In Part A, you will determine the number of moles of copper produced in a reaction between copper oxide and zinc metal. From measurements of the mass of the copper, you will be able to calculate the molar mass of the element copper.

In Part B, you will determine the number of moles of barium sulfate produced in a precipitation reaction between sodium sulfate and barium chloride. From measurements of the mass of the dried barium sulfate, you will be able to calculate the molar mass of the compound barium sulfate.

AIM

- To determine the molar mass of an element, copper, and a compound, barium sulfate

PART A • MOLAR MASS OF COPPER

As directed by your teacher, complete the risk assessment and management table by referring to the safety data sheets (SDSs) or your teacher's risk assessment for the activity.

PRE-LAB SAFETY INFORMATION		
Material used	**Hazard**	**Control**
copper(II) oxide, CuO		
zinc granules		
2 M H_2SO_4		
zinc oxide		

Please indicate that you have understood the information in the safety table.

Name (print): ____________________

I understand the safety information (signature): ____________________

METHOD

1 ▪ Put on your PPE according to the risk assessment and management table you filled in. As relevant, ensure your lab coat is buttoned, your gloves are fitted correctly and comfortably, and your safety glasses/goggles are fitted over your eyes.

2 ▪ Using a weighing bottle/weight boat, find the exact mass of approximately 2 g of CuO. Record this in Table 1.

3 ▪ Transfer the copper oxide to a 250 mL beaker. Add 50 mL of 2 M H_2SO_4. If necessary, heat the mixture by standing the 250 mL beaker in a large beaker of hot water. Stir with a glass rod until the copper oxide is no longer visible.

4 ▪ When all the copper oxide has reacted, add 50 mL water and approximately 3 g of granulated zinc. Swirl and allow to react overnight.

ISBN 978 0 6557 0015 9

5 ▪ If the mixture is still blue at the next lesson, add another 1 g of granulated zinc and allow it to stand for several hours.
6 ▪ Decant the supernatant liquid. Carefully wash the deposited copper with about 50 mL of water. Wash it three more times.
7 ▪ Find the mass of a clean, dry evaporating basin. Record this in Table 1.
8 ▪ Using the minimum amount of water, transfer the copper from the beaker to the evaporating basin.
9 ▪ Dry the copper by heating it gently over a Bunsen burner or similar. When cool, find the mass of the evaporating basin plus copper. Record this in Table 1.
10 ▪ Spray and wipe down your bench, return used equipment, wash equipment as necessary, and remove PPE as directed by your teacher.

RESULTS

Table 1 Mass results

Substance	Mass (g)
copper(II) oxide	
evaporating dish	
evaporating dish and copper	

DISCUSSION

1 Find the mass of copper that was in the copper oxide.

2 Write a balanced equation for the reaction between copper oxide and zinc metal. The products are zinc oxide and copper metal. You do not need to include states.

3 Use your starting mass of copper oxide to calculate the amount, in mol, of copper oxide that reacted.

4 The amount, in mol, of copper produced will be equal to the amount, in mol, of copper oxide that reacted. Write the amount, in mol, of copper produced.

5 Use the amount, in mol, and the mass of copper to calculate a molar mass of copper.

 ISBN 978 0 6557 0015 9

6 According to the periodic table, the molar mass of copper gas is 63.5 g mol^{-1}. How accurate was your experimentally determined value? Suggest two sources of experimental error that may have occurred during your experiment.

PART B • MOLAR MASS OF BARIUM SULFATE

As directed by your teacher, complete the pre-lab safety information by referring to the safety data sheets (SDSs) or your teacher's risk assessment for the activity.

PRE-LAB SAFETY INFORMATION

Material used	Hazard	Control
solid sodium sulfate, Na_2SO_4		
0.5 M barium chloride, $BaCl_2(aq)$		
2 M hydrochloric acid, HCl(aq)		
solid barium sulfate, $BaSO_4$		
heating of solution and crucible		

Please indicate that you have understood the information in the safety table.

Name (print):

I understand the safety information (signature):

MATERIALS

- 0.5 g solid sodium sulfate, Na_2SO_4
- 20 mL of 0.5 M barium chloride, $BaCl_2$
- 3 mL of 2 M hydrochloric acid, HCl
- 200 mL deionised water
- 20 mL warm deionised water
- 2 × 100 mL glass beakers
- 600 mL glass beaker
- 10 mL glass measuring cylinder
- 50 mL burette and stand
- filter funnel
- filter paper
- vacuum flask and vacuum pump (water-jet type)
- glass filter crucible (No. 4 porosity) and rubber adaptor
- glass stirring rod
- wash bottle containing deionised water
- Bunsen burner, tripod and gauze mat
- bench mat
- electronic balance (3 decimal places)
- oven
- safety glasses/goggles
- disposable gloves

METHOD

1 ▪ Put on your PPE according to the risk assessment and management table you filled in. As relevant, ensure your lab coat is buttoned, your gloves are fitted correctly and comfortably, and your safety glasses/goggles are fitted over your eyes.

2 ▪ Accurately and quantitatively weigh out about 0.5 g of Na_2SO_4 and record the mass Table 2.

3 ▪ Add the Na_2SO_4 to 50 mL of deionised water in a 600 mL beaker and stir to dissolve as much of the sample as possible.

4 ▪ Add about 3 mL of 2 M HCl to the solution of Na_2SO_4 and add more water so that the total volume is about 200 mL. Heat the solution until it boils.

5 ▪ Add 15 mL of 0.5 M $BaCl_2$ solution drop by drop from a burette to the hot solution, stirring continuously. A white precipitate of barium sulfate will form.

6 ▪ Boil the mixture for a further minute, then remove it from the heat and allow the precipitate to settle. Ensure no sulfate ions remain in the solution by adding several more drops of $BaCl_2$ solution. If more precipitate forms, add 3 mL of $BaCl_2$ solution and test again for unreacted sulfate ions. If desired, the mixture can be left to stand overnight at this stage.

7 ▪ Weigh a glass filter crucible. Record this in Table 2.

8 ▪ Collect the precipitate in the glass filter crucible using gentle vacuum filtration. (Filtration is faster if most of the liquid is filtered before the bulk of the precipitate is collected in the crucible.) Use about 10 mL of warm deionised water to wash any precipitate remaining in the beaker into the crucible.

9 ▪ Place the crucible and contents in an oven heated to 100–110°C and leave overnight.

10 ▪ Weigh the crucible and contents and record the mass in Table 2.

11 ▪ Spray and wipe down your bench, return used equipment, wash equipment as necessary, and remove PPE as directed by your teacher.

RESULTS

Table 2 Mass results

Substance	Mass (g)
sodium sulfate	
crucible	
crucible and barium sulfate	

DISCUSSION

1 Calculate the mass of the barium sulfate precipitate collected.

2 Write a balanced equation for the reaction between barium chloride and sodium sulfate. The products are barium sulfate and sodium chloride.

3 Calculate the amount, in mol, of sodium sulfate that reacted.

4 The amount, in mol, of barium sulfate produced is equal to the amount, in mol, of sodium sulfate that reacted. Write the amount, in mol, of barium sulfate produced.

5 Use the amount, in mol, and the mass of barium sulfate to calculate the molar mass of barium sulfate.

6 Use a periodic table to find the theoretical molar mass of barium sulfate. Comment on the accuracy of your result.

 ISBN 978 0 6557 0015 9

7 How could you determine the repeatability and reproducibility of the measurements you made in this investigation?

CONCLUSION

PRACTICAL ACTIVITY 10

Chemical composition of a compound

SUGGESTED DURATION

- 30 minutes

MATERIALS

- 20–40 cm piece of magnesium ribbon, Mg
- steel wool
- crucible and lid
- pipeclay triangle
- Bunsen burner, bench mat and tripod
- tongs
- electronic balance (2 decimal places)
- safety glasses/goggles

INTRODUCTION

Compounds contain two or more different elements. The elements in a compound are present in a fixed mass ratio. In this experiment, magnesium metal is heated in oxygen to form the compound magnesium oxide. By finding the masses of the original magnesium and the magnesium oxide, the percentage of magnesium in magnesium oxide can be calculated and the empirical formula can be derived.

AIM

- To determine the percentage by mass of magnesium in magnesium oxide
- To derive the empirical formula of magnesium oxide through experimentation

PRE-LAB SAFETY INFORMATION

Material used	Hazard	Control
magnesium, ribbon, Mg	• when in contact with water, releases flammable gases, which may ignite spontaneously • burns with white-hot flame, which emits UV radiation and may cause eye damage	Wear safety glasses/ goggles. Do not look directly at burning magnesium.
hot crucible	• skin burns	Allow the crucible to cool before handling. Take care if handling a hot crucible.

Please indicate that you have understood the information in the safety table.

Name (print): ____________________

I understand the safety information (signature): ____________________

METHOD

1. Put on your PPE, ensuring your safety glasses/goggles are fitted over your eyes.
2. Set up your equipment as shown in Figure 1.2.7.
3. Clean the 20–40 cm length of magnesium ribbon thoroughly with steel wool.
4. Find the mass of a clean, dry crucible and lid. Coil the magnesium loosely and place it in the crucible. Find the total mass of crucible, lid and magnesium. Record all results in Table 1.
5. Heat the crucible strongly. Using tongs, occasionally lift the lid slowly to allow air to enter the crucible but replace it quickly to avoid loss of magnesium oxide ash.
6. When the reaction appears complete, allow to cool to room temperature and find the total mass of crucible, lid and magnesium oxide.
7. Spray and wipe down your bench, return used equipment, wash equipment as necessary, and remove PPE as directed by your teacher.

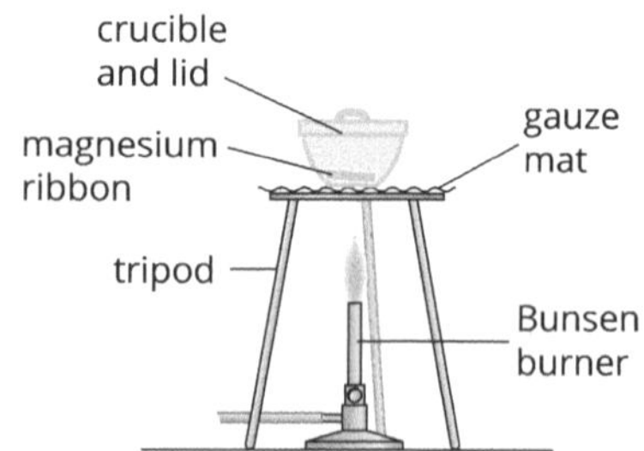

Figure 1.2.7 Experimental set-up

 ISBN 978 0 6557 0015 9

RESULTS

Table 1 Mass results

Material	Mass (g)
empty crucible and lid	
crucible, lid and magnesium	
crucible, lid and magnesium oxide	

DISCUSSION

1 Identify the following variables:

a independent variable

b dependent variable

2 Calculate the mass of magnesium that reacted.

3 Calculate the mass of magnesium oxide that formed.

4 Calculate the percentage by mass of magnesium in magnesium oxide.

5 Calculate the:

a amount, in mol, of magnesium reacted

b mass of oxygen, in g, in your sample of magnesium oxide

c amount, in mol, of oxygen in the sample

d ratio of moles of magnesium and oxygen.

6 From Question 5, derive the empirical formula of magnesium oxide.

7 a Find out the result for percentage by mass obtained by three other groups that had different starting masses of magnesium (Table 2).

Table 2 Group results

Group	Starting mass of magnesium (g)	Percentage by mass of magnesium (%)
your group		
other group 1		
other group 2		
other group 3		

b Are the results the same or different?

c Would you expect them to be the same or different? Why?

8 Percentage composition and empirical formula are both ways of expressing the chemical composition of a compound. Explain the difference between them.

9 The theoretical percentage by mass of magnesium in magnesium oxide is 60.3%. Calculate your percentage error and comment on the accuracy of your data.

10 Describe two random errors that could have affected your result.

11 List some improvements you could make to this experiment to address the errors identified in Question 10.

CONCLUSION

ISBN 978 0 6557 0015 9

PRACTICAL ACTIVITY 11

Investigating hydrocarbons

SUGGESTED DURATION

- 50 minutes

INTRODUCTION

Hydrocarbons are covalently bonded molecules composed only of carbon and hydrogen.

A saturated hydrocarbon has only single bonds between carbon atoms. An unsaturated hydrocarbon contains double or triple bonds between at least two carbon atoms in the hydrocarbon. Permanganate solution only reacts with unsaturated hydrocarbons. The distinctive colour of the permanganate disappears as the double bond in the hydrocarbon breaks and an OH group is added to each of the carbon atoms that was originally involved in the double or triple bond.

$5CH_2{=}CH_2 + 2H_2O + 2MnO_4^- + 6H^+ \rightarrow 5CH_2OHCH_2OH + 2Mn^+$

AIM

- To investigate the properties of different families of hydrocarbons

MATERIALS

- 10 mL of 1 M hydrochloric acid, $HCl(aq)$, in dropper bottle
- 10 mL of 1 M sodium hydroxide, $NaOH(aq)$, in dropper bottle
- 0.002 M potassium permanganate, $KMnO_4(aq)$, in dropper bottle
- 1 M sulfuric acid, $H_2SO_4(aq)$
- dropper bottles containing different hydrocarbons: cyclohexane, cyclohexene, heptane, hexene
- small test tube for each hydrocarbon
- water coloured with red food dye
- safety glasses/goggles
- disposable gloves

PRE-LAB SAFETY INFORMATION

Material used	Hazard	Control
1 M hydrochloric acid, $HCl(aq)$	• non-hazardous < 3 M	Wear safety glasses/goggles.
1 M sodium hydroxide, $NaOH(aq)$	• causes severe skin burns and eye damage	Wear safety glasses/goggles, disposable gloves and a lab coat.
1 M sulfuric acid, $H_2SO_4(aq)$	• causes skin irritation • causes serious eye irritation	Wear safety glasses/goggles, disposable gloves and a lab coat.
0.002 M acidified potassium permanganate, $KMnO_4(aq)$	• harmful to aquatic life with long-lasting effects • readily stains skin and clothes	Wear safety glasses/goggles, disposable gloves and a lab coat.
heptane	• highly flammable liquid and vapour • may be fatal if swallowed and enters airways • causes skin irritation • may cause drowsiness or dizziness • very toxic to aquatic life with long-lasting effects	Wear safety glasses/goggles, disposable gloves and a lab coat. Use in a fume hood.
hex-1-ene	• highly flammable liquid and vapour • may be fatal if swallowed and enters airways	Wear safety glasses/goggles, disposable gloves and a lab coat. Use in a fume hood.
cyclohexane	• highly flammable liquid and vapour • may be fatal if swallowed and enters airways • causes skin irritation • may cause drowsiness or dizziness • very toxic to aquatic life with long-lasting effects	Wear safety glasses/goggles, disposable gloves and a lab coat. Use in a fume hood.
cyclohexene	• highly flammable liquid and vapour • harmful if swallowed • may be fatal if swallowed and enters airways • toxic in contact with skin • very toxic to aquatic life with long-lasting effects	Wear safety glasses/goggles, disposable gloves and a lab coat. Use in a fume hood.

Disposal of waste: Do not pour hydrocarbons down the sink. Place all organic waste in a waste bottle in the fume cupboard.

Please indicate that you have understood the information in the safety table.

Name (print): ______________________

I understand the safety information (signature): ______________________

METHOD

The following tests should be carried out in a fume cupboard. Your teacher may complete this practical activity as a demonstration in a fume cupboard.

Put on your PPE, ensuring your lab coat is buttoned, your gloves are fitted correctly and comfortably, and your safety glasses/goggles are fitted over your eyes.

PART A • SOLUBILITY OF HYDROCARBONS IN WATER

1 ▪ Mix 10 drops of each hydrocarbon to be tested with 10 drops of coloured water in a small test tube. Record your observations about the solubility of each hydrocarbon in Table 1.

PART B • REACTIVITY OF HYDROCARBONS WITH ACIDS AND BASES

2 ▪ Place 10 drops of each hydrocarbon to be tested in a small test tube and add 10 drops of 1 M hydrochloric acid. Use a different test tube for each hydrocarbon. Record your observations in Table 1.

3 ▪ Place 10 drops of each hydrocarbon in a small test tube and add the same number of drops of 1 M sodium hydroxide. Use a different test tube for each hydrocarbon.

PART C • SATURATION AND UNSATURATION IN HYDROCARBONS

4 ▪ Place 10 drops of each hydrocarbon to be tested in a small test tube. Add two drops of potassium permanganate solution dropwise to each test tube and then add one drop of sulfuric acid. Note whether the distinctive colour of the permanganate solution remains or disappears.

5 ▪ Spray and wipe down your bench, return used equipment, wash equipment as necessary, and remove PPE as directed by your teacher.

RESULTS

Table 1 Solubility and reactivity

Hydrocarbon	Solubility in water	Reactivity with 1 M HCl	Reactivity with 1 M NaOH	Reaction with permanganate solution
heptane				
hex-1-ene				
cyclohexane				
cyclohexene				

DISCUSSION

1 State whether any of the hydrocarbons were soluble in water. Deduce what this tells you about their attraction to water.

2 a In the past, bromine solution was used to test whether a hydrocarbon was saturated or unsaturated. Bromine is a reddish liquid at room temperature. It is also highly toxic. Write a chemical equation for the reaction that would occur between the unsaturated hydrocarbon(s) used in this experiment and bromine solution ($Br_2(aq)$).

 ISBN 978 0 6557 0015 9

b Suggest why the experiment using bromine solution worked particularly well.

3 Summarise the similarities you observed in the different hydrocarbons with regard to their solubility in water and reactivity with acids and bases.

4 Classify each of the hydrocarbons you investigated in this practical activity as saturated or unsaturated.

5 Evaluate whether this practical activity allowed effective investigation of the properties of hydrocarbons.

6 Identify any improvements you could make to this practical activity and/or how you could further extend the investigation.

CONCLUSION

PRACTICAL ACTIVITY 12

Modelling functional groups

SUGGESTED DURATION

- 50 minutes

MATERIALS

- molecular model building kit or golf-ball-sized lumps of different colours of modelling dough and different-length straws

AIM

- To examine the bonding, shape and nomenclature of a number of organic molecules with common functional groups
- To investigate the concept of structural isomers

PART A • FUNCTIONAL GROUPS

METHOD

Consider each molecule in Table 1.

1. Draw the structural formula.
2. Construct a three-dimensional model with the modelling kit.
3. Sketch the three-dimensional model.
4. Write the semi-structural formula.
5. Return used equipment.

RESULTS

Table 1 Part A results

Name	Structural formula	Three-dimensional model	Semi-structural formula
methane			
ethane			
propane			
ethene			
but-1-ene			

continues over

 ISBN 978 0 6557 0015 9

Table 1 Part A results (continued)			
Name	**Structural formula**	**Three-dimensional model**	**Semi-structural formula**
but-2-ene			
chloroethane			
1-chloropropane			
2-chloropropane			
methanol			
ethanol			
butan-2-ol			
methanoic acid			
ethanoic acid			

PART B • ISOMERS

METHOD

Consider each molecular formula in Table 2.

1. Construct a model of all possible isomers.
2. Draw a structural formula of each isomer
3. Provide the systematic name of each isomer.

RESULTS

Table 2 Part B results

Molecular formula	Structural formula and name of each isomer
C_5H_{12}	
C_4H_9OH	

DISCUSSION

1 What type of bonding holds the atoms together within each molecule that you made?

2 Name the different functional groups present within each group of molecules constructed in Part A.

3 With reference to the molecules you constructed, describe structural isomers.

4 What factors limit the number of structural isomers a molecule can have?

CONCLUSION

 ISBN 978 0 6557 0015 9

PRACTICAL ACTIVITY 13

Investigating properties of slime, an addition polymer

SUGGESTED DURATION

- 20 minutes

INTRODUCTION

Polyvinyl alcohol is an addition polymer formed from ethenol (vinyl alcohol). The polymer can dissolve in water because the –OH groups along the polymer chain form hydrogen bonds with water molecules (Figure 1.2.8).

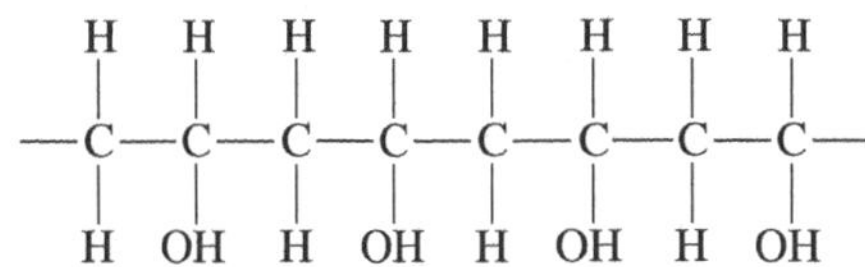

Figure 1.2.8 A section of PVA chain

Addition of borax creates cross-links between the chains. Borate ions, $B(OH)_4^-$, present in borax solution form four hydrogen bonds between two polymer chains, trapping water molecules (Figure 1.2.9). The result is a gel that is more than 90% water.

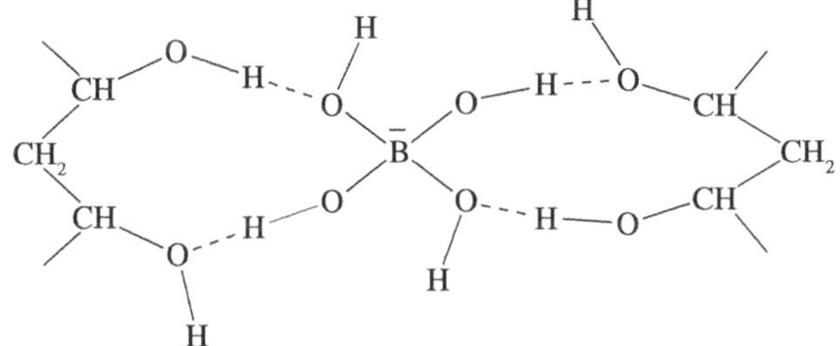

Figure 1.2.9 Borate ions cross-linking two PVA polymer chains

Unlike cross-links formed by covalent bonds, the links in the gel break and re-form easily, giving the material interesting properties. The gel behaves as a non-Newtonian fluid: it flows if pulled gently and snaps if pulled sharply. A lump of slime placed in a container will slowly flow to adopt the shape of the container.

MATERIALS

- 60 mL of 6% polyvinyl alcohol
- food colouring (various colours)
- 10 mL of 4% borax (sodium tetraborate) solution
- paper cups
- wooden ice-cream stirring sticks
- zip-lock bag
- safety glasses/goggles
- disposable gloves

AIM

- To observe the changes in the physical properties of a polymer as a result of the effect of weak cross-linking (hydrogen bonds) between chains
- To observe a polymer with unusual properties

PRE-LAB SAFETY INFORMATION		
Material used	**Hazard**	**Control**
6% polyvinyl alcohol	• non-hazardous at all concentrations	
food colouring	• stains skin and clothes	Wear disposable gloves and lab coat.
4% borax (sodium borate) solution	• higher concentrations may damage fertility or an unborn child—take care even with lower concentrations	Wear safety glasses/goggles, disposable gloves and skin protection.

Please indicate that you have understood the information in the safety table.

Name (print): ______________________

I understand the safety information (signature): ______________________

METHOD

1 ▪ Put on your PPE, ensuring your lab coat is buttoned, your gloves are fitted correctly and comfortably, and your safety glasses/goggles are fitted over your eyes.
2 ▪ Place 60 mL of 6% PVA solution in a paper cup.
3 ▪ Add a few drops of food colouring and stir with a wooden stick.
4 ▪ Add 10 mL of 4% borax solution. Begin stirring immediately.
5 ▪ When the slime has formed, take some in your gloved hand and slowly stretch it. Record your observations in Table 1.
6 ▪ Repeat step 5 but very quickly. Do you notice any differences in how the slime behaves? Record your observations in Table 1.
7 ▪ Spray and wipe down your bench, return used equipment, wash equipment as necessary, and remove PPE as directed by your teacher.

RESULTS

Table 1 Observations of the slime

Method of handling the slime	Observations
slowly stretching the slime	
stretching the slime very quickly	

DISCUSSION

1 Identify the physical properties that change as a result of the addition of borax to the polyvinyl alcohol. Why do they change?

2 Predict what would be the effect of adding more borax to the PVA.

3 PVA is a polymer formed from addition polymerisation. What is addition polymerisation?

4 Explain why PVA is soluble in water.

 ISBN 978 0 6557 0015 9

5 PVA is used to make hospital laundry bags. What properties make PVA suitable for this application?

6 Identify the repeating unit in Figure 1.2.8. Draw the monomer used to make PVA.

7 How would the physical properties of a polymer cross-linked by covalent bonds differ from this slime, which is cross-linked by hydrogen bonds only?

CONCLUSION

ISBN 978 0 6557 0015 9

PRACTICAL ACTIVITY 14

Making a bioplastic

SUGGESTED DURATION

- Part A = 25 minutes, Part B = 40 minutes (plus time for observations after drying)

INTRODUCTION

Bioplastics are manufactured from raw materials that are sourced from plants. In this experiment, you will make a plastic from starch extracted from potatoes. Starch is a polymer made of long chains of glucose joined together. When starch is dried from an aqueous solution, the hydrogen bonding between chains causes a film to form.

Propane-1,2,3-triol, also known as glycerol or glycerine, is a plasticiser. It has three hydroxyl groups which allow it to attract water. The water can get in amongst starch chains preventing brittleness and resulting in more 'plastic' properties.

AIM

- To make a plastic from potato starch
- To investigate the effect of adding a 'plasticiser' on the properties of the polymer

PRE-LAB SAFETY INFORMATION

Material used	Hazard	Control
0.1 M HCl	• non-hazardous < 3 M	Wear safety glasses/goggles.
0.1 M NaOH	• non-hazardous	Wear safety glasses/goggles.
food colouring	• readily stains skin and clothes	Wear disposable gloves and a lab coat.
propane-1,2,3-triol	• non-hazardous	

Please indicate that you have understood the information in the safety table.

Name (print): ______________________

I understand the safety information (signature): ______________________

MATERIALS

- 10 mL of 0.1 M hydrochloric acid, HCl
- 10 mL of 0.1 M sodium hydroxide, NaOH
- 100 g potato
- 500 mL deionised water
- food colouring
- 2 mL propane-1,2,3-triol (glycerol/glycerine)
- grater
- 500 mL glass beaker
- large mortar and pestle
- 250 mL glass beaker
- large watch glass
- Bunsen burner, bench mat, tripod and gauze mat
- glass stirring rod
- 2 white tiles
- universal indicator paper
- plastic Pasteur pipette
- 25 mL glass measuring cylinder
- 10 mL glass measuring cylinder
- tea strainer
- electronic balance
- safety glasses/goggles
- disposable gloves

METHOD

PART A • EXTRACTING STARCH FROM THE POTATO

1. Put on your PPE, ensuring your lab coat is buttoned, your gloves are fitted correctly and comfortably, and your safety glasses/goggles are fitted over your eyes.
2. Grate about 100 g of potato. The potato does not need to be peeled, but it should be clean. Place the grated potato into the mortar.
3. Add about 100 mL of deionised water to the mortar, and grind the potato carefully.
4. Pour the liquid off the potato mixture through the tea strainer into the 500 mL beaker, leaving the potato behind in the mortar.
5. Repeat steps 3 and 4 twice more.
6. Leave the mixture to settle in the beaker for 5 minutes.
7. Decant the water from the beaker, leaving behind the white starch, which should have settled in the bottom. Add about 100 mL of distilled water to the starch and stir gently. Leave to settle again for 5 minutes and then decant the water again, leaving the starch behind.

PART B • MAKING THE PLASTIC FILM

1. Add 22 mL of deionised water to the 250 mL beaker and add 4 g of the potato starch collected in Part A.
2. Add 3 mL of hydrochloric acid and 2 mL of propane-1,2,3-triol to the water and potato starch mixture. Stir it gently.

ISBN 978 0 6557 0015 9

3 ▪ Place the watch glass on the beaker and heat the mixture over the Bunsen burner. Bring the mixture carefully to the boil and then boil gently for 15 minutes. Do not boil it dry. If the mixture looks as though it is completely drying out, stop heating.
4 ▪ Dip the glass rod into the mixture and dot it onto the indicator paper to measure the pH. Add enough sodium hydroxide solution to neutralise the mixture, testing after each addition with indicator paper. You will probably need to add about the same amount as you did of acid at the beginning (3 mL).
5 ▪ Add a drop of food colouring and mix thoroughly.
6 ▪ Pour the mixture onto the white tile and spread it out with the glass rod so there is an even covering over the tile.
7 ▪ Repeat steps 1–6 to produce a second plastic film, but this time leave out the propane-1,2,3-triol and boil for 10 minutes instead of 15.
8 ▪ Label the mixtures and leave them to dry completely. It takes about one day on a sunny windowsill, or two days at room temperature. Alternatively, use a drying cabinet, which takes about 90 minutes at 100°C.
9 ▪ When your polymers are dry, investigate their strength and flexibility and record your observations in Table 1.
10 ▪ Spray and wipe down your bench, return used equipment, wash equipment as necessary, and remove PPE as directed by your teacher.

RESULTS

Table 1 Observations of bioplastics

Type of bioplastic	Observations
made with propane-1,2,3-triol	
made without propane-1,2,3-triol	

DISCUSSION

1 Compare the properties of the two plastics you made. What is the effect of the plasticiser on the properties of the polymer?

2 Starch is a polymer formed from condensation polymerisation of glucose molecule monomers. What occurs during condensation polymerisation?

3 Would you expect this polymer to be compostable? Explain your answer.

4 Find out ways that starch polymers from potato are used in society. For each application, identify the characteristics of the polymer that make it suitable for that use.

CONCLUSION

ISBN 978 0 6557 0015 9

EXAM-STYLE QUESTIONS

Multiple-choice questions

Question 1

Identify the sample that contains the most atoms.

A. 10.0 g of CO

B. 10.0 g of CO_2

C. 10.0 g of SO_3

D. 10.0 g of NO_2

Question 2

Determine the mass of a single sulfur atom.

A. $\frac{6.02 \times 10^{23}}{32.1}$

B. $6.02 \times 10^{23} \times 32.1$

C. $\frac{32.1}{6.02 \times 10^{23}}$

D. $\frac{1}{6.02 \times 10^{23}}$

Question 3

Select the compound that contains the greatest percentage by mass of sulfur.

A. SO_3

B. H_2SO_4

C. $MgSO_4$

D. $H_2S_2O_7$

Question 4

Identify the hydrocarbon molecule most likely to react with permanganate.

A. ethane because it is saturated

B. ethene because it is saturated

C. ethane because it is unsaturated

D. ethene because it is unsaturated

Question 5

Identify the molecule that will have the highest melting point.

A. propane

B. propan-1-ol

C. 1-chloropropane

D. propanoic acid

Question 6

What is the correct IUPAC name of the molecule with the semi-structural formula $CH_3CH_2CHOHCH_3$?

A. butanoic acid

B. butan-1-ol

C. butan-2-ol

D. propan-2-ol

ISBN 978 0 6557 0015 9

EXAM-STYLE QUESTIONS

Question 7

How many structural isomers exist with the molecular formula C_4H_8?

A. 1

B. 2

C. 3

D. 4

Question 8

The following represents a section of a polymethyl methacrylate (perspex) polymer.

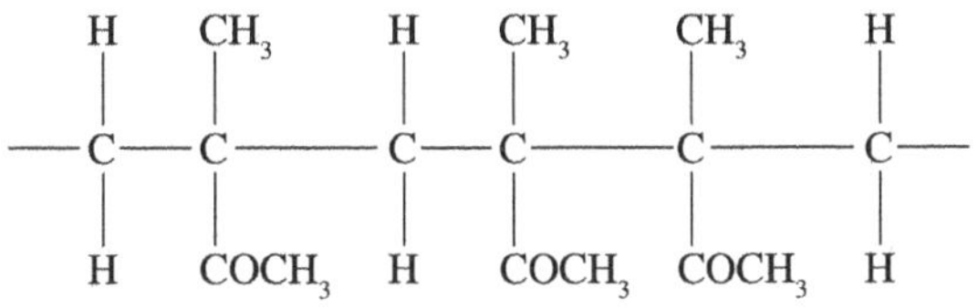

```
   H   CH3      H   CH3     CH3       H
   |   |        |   |       |         |
 --C---C--------C---C-------C---------C--
   |   |        |   |       |         |
   H   COCH3    H   COCH3   COCH3     H
```

Which option correctly describes the way this polymer was produced?

A. addition polymerisation, with no other products formed

B. addition polymerisation, with another small product also formed

C. condensation polymerisation, with no other products formed

D. condensation polymerisation, with another small product also formed

Question 9

The properties of a polymer can be modified by altering the amount of branching present. A polymer with a high amount of branching is likely to be:

A. more flexible, with a higher melting point

B. more flexible, with a lower melting point

C. less flexible, with a higher melting point

D. less flexible, with a lower melting point

Question 10

A segment of a polymer molecule is drawn below.

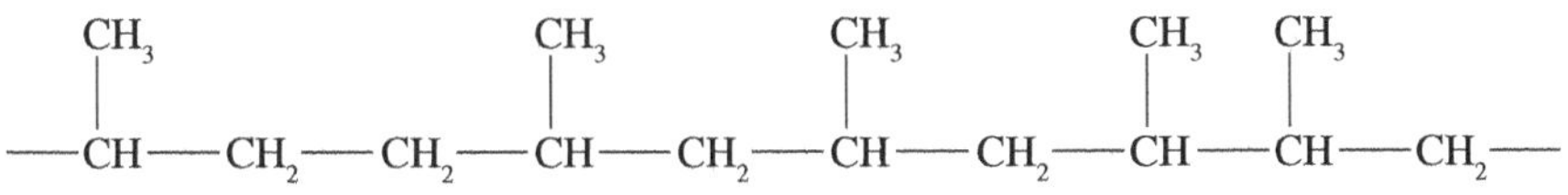

```
  CH3                   CH3         CH3          CH3    CH3
  |                     |           |            |      |
--CH---CH2---CH2---CH---CH2---CH---CH2---CH---CH---CH2--
```

The monomer(s) used to form this polymer is (are):

A. propene

B. ethene and ethane

C. but-2-ene

D. methane and ethene

EXAM-STYLE QUESTIONS

Short-answer questions

Question 1

The following mass spectrum shows the isotopic pattern of a particular sample of rubidium.

Isotopes of rubidium (Rb)

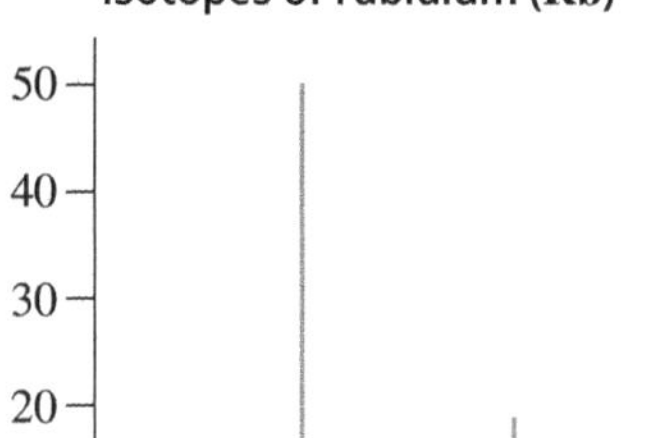

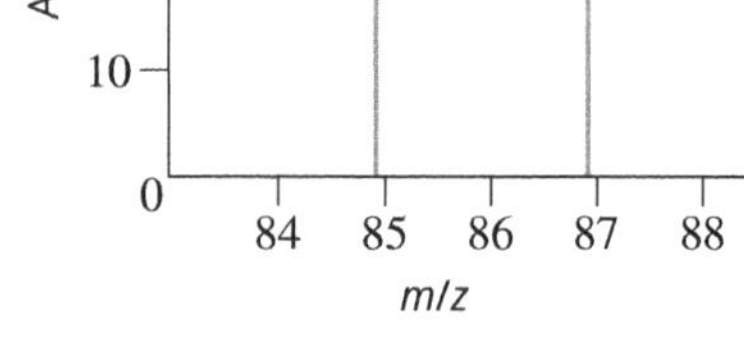

a. What is the definition of an isotope? 2 marks

b. How many isotopes are evident in the sample? 1 mark

c. Use the information in the spectrum to calculate the relative atomic mass of rubidium. 3 marks

Question 2

A 19.0 g sample of fluorine gas reacts exactly with another element to form 39.0 g of XF_2. Use the mass values given to determine the identity of the element X. 3 marks

Question 3

a. Draw the structural formula of molecule of butanoic acid. 1 mark

 ISBN 978 0 6557 0015 9

b. Circle and name the functional group present in butanoic acid. 2 marks

c. Draw the structural formula and name a structural isomer of butanoic acid that is also a carboxylic acid. 2 marks

Question 4

The properties of most smaller carbon compounds should be predictable from their structure. Scientists understand the effect of particular functional groups and molecule length on the properties of the molecule. Use your knowledge of chemical structure to answer the following questions.

a. Rank the following molecules in order of increasing boiling point. 1 mark

butan-1-ol   butane   1-chlorobutane   butanoic acid

b. Rank the following in order of increasing solubility in water. 1 mark

pentan-1-ol   ethanol   propane   1-chloropentane

c. When comparing the boiling points of 1-chloropropane and propan-1-ol, which type of bonding is the main consideration? 1 mark

Question 5

Polyethene is produced in large quantities by addition polymerisation of ethene.

a. Draw a structural formula for ethene. 1 mark

b. What is meant by the term 'addition polymerisation'? 1 mark

c. One particular strand of polyethene has a molar mass of 44 800 $g\,mol^{-1}$. How many monomer molecules polymerised to form that chain? 1 mark

d. Polyethene can form as a high-density or a low-density product. Describe how the structures and melting points of these two materials differ and explain the reasons for the difference. 3 marks

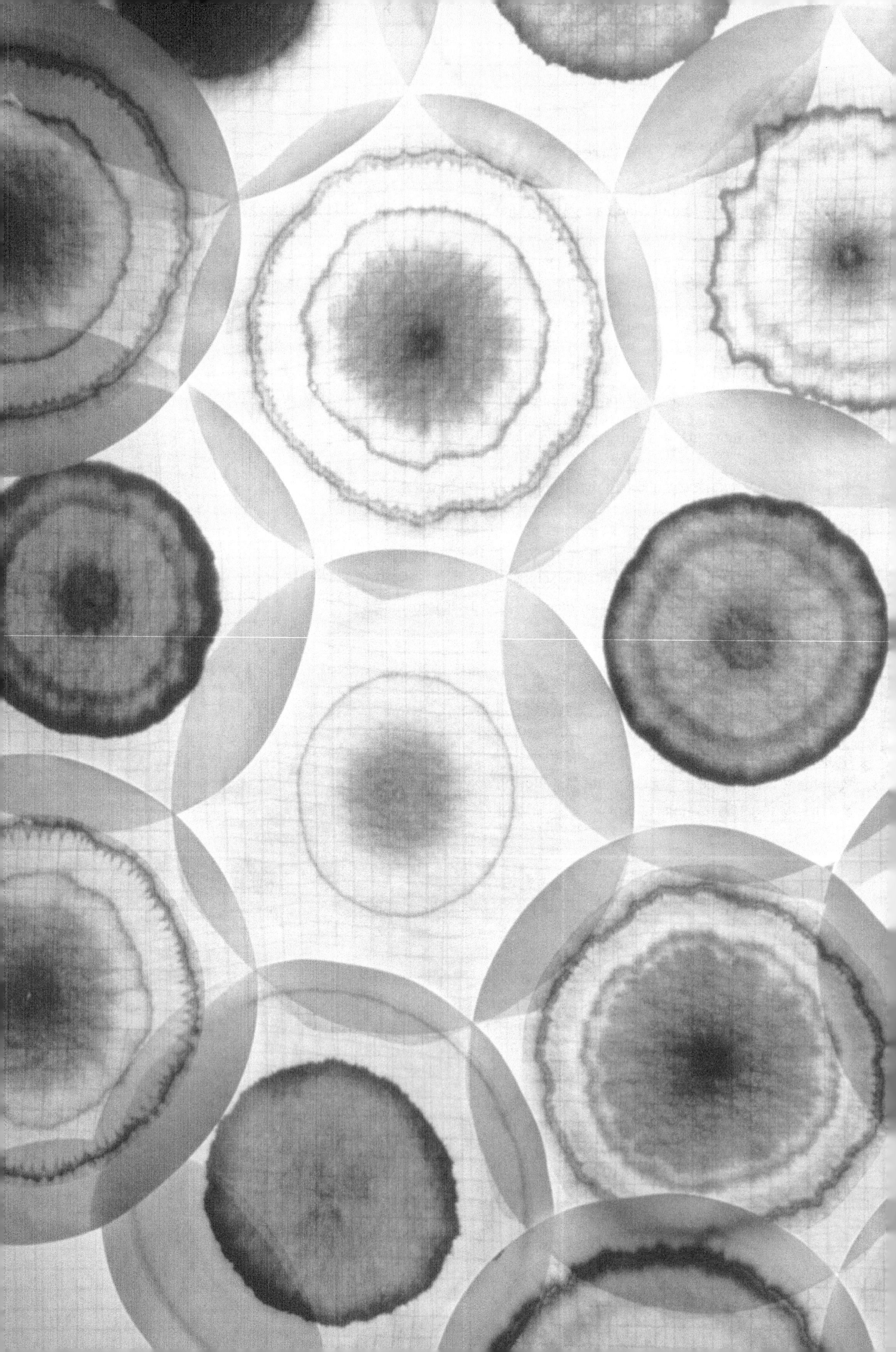

UNIT 1

How can the diversity of materials be explained?

AREA OF STUDY 3

How can chemical principles be applied to create a more sustainable future?

Outcome

On completion of this unit the student should be able to investigate and explain how chemical knowledge is used to create a more sustainable future in relation to the production or use of a selected material.

Key knowledge

Scientific evidence

- the distinction between primary and secondary data
- the nature of evidence and information: distinction between opinion, anecdote and evidence; and scientific and non-scientific ideas
- the quality of evidence, including validity and authority of data and sources of possible errors or bias
- methods of organising, analysing and evaluating secondary data
- the use of a logbook to authenticate collated data

Sustainability

- sustainability concepts and principles: green chemistry principles, sustainable development, and the transition from a linear economy towards a circular economy
- identification of sustainability concepts and principles relevant to the selected research question

Scientific communication

- chemical concepts specific to the investigation: definitions of key terms; and use of appropriate chemical terminology, conventions and representations
- characteristics of effective science communication: accuracy of chemical information; clarity of explanation of chemical concepts, ideas and models; contextual clarity with reference to importance and implications of findings; conciseness and coherence; and appropriateness for purpose and audience
- the use of data representations, models and theories in organising and explaining observed phenomena and chemical concepts, and their limitations
- the influence of social, economic, legal and/or political factors relevant to the selected research question
- conventions for referencing and acknowledging sources of information

Investigating the issue of banning single-use plastics in Victoria

You will undertake an investigation involving the selection and evaluation of a recent discovery, innovation, advance, case study, issue or challenge linked to the knowledge and skills developed in Unit 1 Area of Study 1 and/or Area of Study 2, including consideration of sustainability concepts (green chemistry principles, sustainable development and the transition towards a circular economy).

ASSESSMENT FOR OUTCOME 3

A response to a question involving the production or use of a selected material, including reference to sustainability.

SUGGESTED DURATION

A minimum of seven hours of class time should be devoted to undertaking and communicating findings related to Area of Study 3. As per Areas of Study 1 and 2, your teacher will be your primary guide. However, a suggested time frame is set out in the table below for your convenience. An introduction and guide to a sample investigation of a research question is provided after the table. This sample investigation focuses on single-use plastics and whether they should be banned in Victoria.

Refer to the Toolkit at the beginning of this book for more detailed information on designing and conducting scientific investigations and presenting a scientific report.

APPROACHING YOUR INVESTIGATION

Scientific and technological research has provided modern society with a diverse range of useful chemicals, materials and products. Recognition of harmful effects of the manufacture and/or disposal of some of these materials has led people to demand that sustainability be considered alongside cost and benefits of the materials.

Whichever topic you choose for your investigation, you must consider the following sustainability concepts.

- The seven *green chemistry principles* that have been identified as being particularly relevant to the study of VCE Chemistry:
  - Atom economy
  - Catalysis
  - Design for degradation
  - Design for energy efficiency
  - Designing safer chemicals
  - Prevention of wastes
  - Use of renewable feedstocks
- *Sustainable development*, which provides guidelines for meeting the needs of present generations without compromising the ability of future generations to meet their needs. The United Nations Sustainable Development Goal 12 mandates responsible consumption and production and is particularly relevant to this investigation; others may also be relevant and should be considered
- *Transition towards a circular economy* in which resources are used to make materials and are then recycled, composted or enriched for re-use rather than simply being disposed of.

SAMPLE INVESTIGATION: EXPLORING AN ISSUE

One approach you could take is to identify and investigate an issue. In this sample investigation, the issue is the banning of single-use plastics in Victoria. It relates to sustainability concepts of:

- prevention of waste
- use of renewable feedstocks
- sustainable development mandate of responsible consumption
- transition to a circular economy.

Single-use plastics have long provided cheap and convenient packaging for food and water as well as supplies for medical equipment and scientific research, such as gloves, syringes and sample bags. Plastic is suitable for these applications because of its low cost, ease of production, strength and resilience. However, public concern has grown alongside awareness of the impact of plastic pollution on the environment. Many countries have banned single-use plastics for sustainability reasons; however, there are concerns about what this will mean for individuals and industries that rely on single-use plastics.

In February 2021, the Victorian Government announced a ban on specific items currently made from single-use plastics. Although many people have welcomed this, some people are concerned that it will have a flow-on cost for consumers. Others are concerned the ban does not go far enough.

Research investigation section	Key step	Suggested time allocation (min)
Planning your research investigation	Step 1: Developing your ideas related to sustainability concepts	30–60
	Step 2: Choosing your investigation topic and developing your draft research questions	30
Conducting your investigation and collating secondary data	Step 3: Conducting your investigation	90–120
	Step 4: Refining your investigation and redrafting your research question	10
	Step 5: Summarising secondary data and main points	60
	Step 6: Analysing and evaluating the sustainability of your material	60–90
Communicating your findings	Step 7: Planning the structure of your response	30
	Step 8: Completing your response	120–180

ISBN 978 0 6557 0015 9

GUIDING QUESTIONS

You need to develop guiding questions to break down the research question into more manageable chunks for which you should be able to research answers. It is expected you would develop further lines of inquiry as you carry out research and refine your research question.

Research question: *Should the sale of single-use plastics be banned in Victoria?*

Sample guiding questions:

1 What are the main applications of single-use plastics in Victoria?
2 What are the problems with using single-use plastics in Victoria?
3 What type of polymers are currently used for single-use plastics?
4 What are the raw materials currently used for manufacturing these plastics and how are they sourced?
5 Are there alternative, 'greener', raw materials that could be used to make these plastics?
7 What are the current options for recovering and/or recycling these plastics?
8 What percentage of these plastics are currently recovered or recycled? How could this be improved?
9 Are there future alternatives of recovering and/or recycling of these plastics that are in development?

INVESTIGATING THE ISSUE

An example of a source of secondary data that can be used to investigate the issue is the following extract from information placed on the Victorian Government website in 2021. The information announces a ban on certain single-use plastics from 2023. It is written by the government that is legislating the ban and is very supportive of the ban. As you locate and use sources of information, you could complete a research table such as in the Toolkit on pages xx–xxi.

SAMPLE SECONDARY SOURCE: SINGLE-USE PLASTICS BAN

Plastic pollution harms our health, wildlife, and the environment

These problematic single-use plastics will be banned across Victoria by February 2023.

Single-use straws, cutlery, plates, drink stirrers, expanded polystyrene food and drink containers, and cotton bud sticks will be banned from sale or supply in Victoria by February 2023.

This ban will reduce plastic pollution.

Single-use plastics:

- make up a third of the litter we see on our streets and in our waterways—they are costly to clean up and difficult to recycle
- are often used for only a few minutes but remain in the environment for a long time
- harm the environment—they break down into microplastics which harm wildlife and contaminate our food and water.

What exemptions will exist?

Exemptions will be considered for scientific, medical or forensic purposes or facilities that require the continued use of banned items for health and safety reasons, where no alternatives are available.

Single-use plastic straws will remain available to people who need them for health or disability reasons.

What can be used instead of single-use plastic items?

Avoid single-use plastics where you can and choose reusables:

- straws—drink from a glass or cup, or choose a reusable straw made from bamboo, stainless steel or silicone
- cutlery—dine in or eat finger food, or choose reusable cutlery made from stainless steel, heavyweight plastic or bamboo
- plates—dine in or eat finger food, or choose reusable plates made from ceramic or heavyweight plastic
- drink stirrers—choose a reusable metal teaspoon or swizzle stick
- food and drink containers—BYO reusable glass, heavyweight plastic or metal containers
- cotton buds—choose cotton pads or buds with bamboo sticks.

If you cannot avoid or use reusable items, then choose non-plastic materials such as FSC® certified paper or wood, bamboo or recycled aluminium.

Source: The Victorian Government Single-use plastics ban web page

Source: Victorian Government. (n.d.). *Single-use plastics ban.* https://www.vic.gov.au/single-use-plastics

Relevant question	Summary of information
What are the main applications of single-use plastics in Victoria?	straws, cutlery, plates, drink stirrers, expanded polystyrene food and drink containers, and cotton bud sticks
What are the problems with using plastic water bottles and single-use plastics in Victoria?	litter, difficult to recycle, long-lived, microplastics in environment
Are there alternative, 'greener', raw materials that could be used to make these plastics?	There are greener alternatives for making products such as bamboo, stainless steel, silicone, wood, recycled aluminium Source doesn't mention an alternative raw material for plastics

After reading this source, some more guiding questions might be developed.

10 What are microplastics?
11 What is the average lifespan for single-use plastics?
12 What are environmental impacts of suggested alternatives to plastics?

RESPONSE TO THE INVESTIGATION

There are different ways you could provide a response to the investigation and this may be guided by your teacher. In this example, you are required to write an editorial for a newspaper in which you outline the sustainability of the use of single-use plastics and recommend whether Victoria should ban their use, or go even further. It will be important to summarise the key points of manufacture and disposal/recycling of these plastics and possible alternatives in substantiating your response. You will also need to draw on your evaluation of the sustainability of single-use plastics.

REFERENCES AND ACKNOWLEDGEMENTS

Remember to acknowledge any secondary sources used in the course of the investigation. This may include:

- scientific journals
- medical articles and texts
- internet sources; for example, TED talks
- newspaper articles.

When referencing published sources, follow accepted protocols for presenting bibliographical data, such as APA citation style. For example:

Victorian Government. (n.d.). *Single-use plastics ban.* https://www.vic.gov.au/single-use-plastics

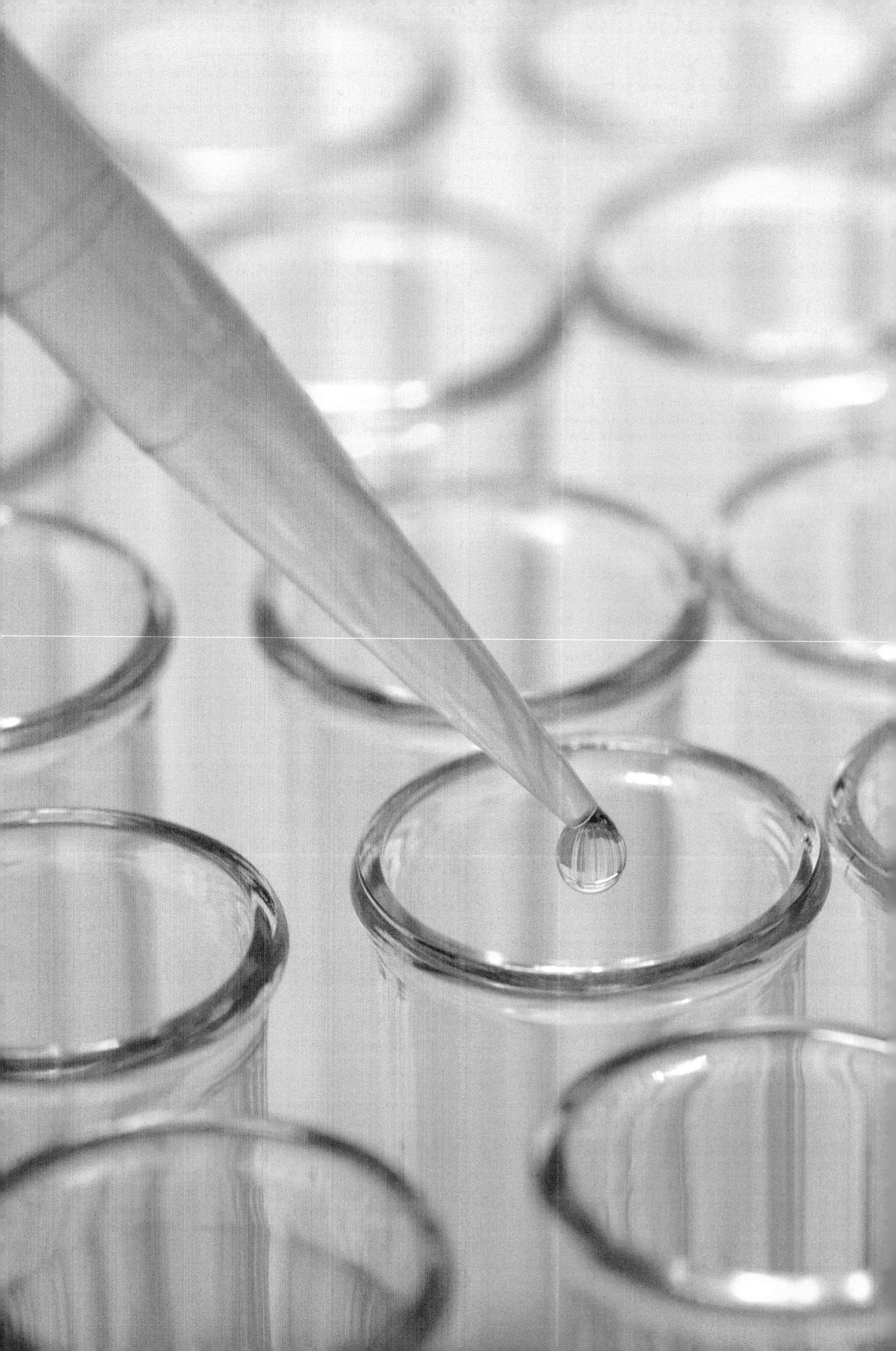

UNIT 2

How do chemical reactions shape the natural world?

AREA OF STUDY 1

How do chemicals interact with water?

Outcome

On completion of this unit the student should be able to explain the properties of water in terms of structure and bonding, and experimentally investigate and analyse applications of acid–base and redox reactions in society.

Key knowledge

Water as a unique chemical

- existence of water in all three states at Earth's surface, including the distribution and proportion of available drinking water
- explanation of the anomalous properties of H_2O (ice and water) with reference to hydrogen bonding:
  - trends in the boiling points of Group 16 hydrides
  - the density of solid ice compared with liquid water at low temperatures
  - specific heat capacity of water including units and symbols
- the relatively high latent heat of vaporisation of water and its impact on the regulation of the temperature of the oceans and aquatic life

Acid–base (proton transfer) reactions

- the Brønsted–Lowry theory of acids and bases, including polyprotic acids and amphiprotic species, and the writing of balanced ionic and full equations, with states, for their reactions in water
- the distinction between strong and weak acids and strong and weak bases, and between concentrated and dilute acids and bases, including common examples
- neutralisation reactions to produce salts:
  - reactions of acids with metal carbonates and hydroxides, including balanced full and ionic equations, with states
  - types of antacids and their use in the neutralisation of stomach acid
- use of the logarithmic pH scale to rank solutions from most acidic to most basic; calculation of pH for strong acid and strong base solutions of known concentration using the ionic product of water (K_w at a given temperature)
- accuracy and precision in measurement as illustrated by the comparison of natural indicators, commercial indicators, and pH meters to determine the relative strengths of acidic and basic solutions
- applications of acid–base reactions in society: for example, natural acidity of rain due to dissolved CO_2 and the distinction between the natural acidity of rain and acid rain, or the action of CO_2 forming a weak acid in oceans and the consequences for shell growth in marine invertebrates

Redox (electron transfer) reactions

- oxidising and reducing agents, and redox reactions, including writing of balanced half and overall redox equations (including in acidic conditions), with states
- the reactivity series of metals and metal displacement reactions, including balanced redox equations, with states
- applications of redox reactions in society: for example, corrosion or the use of simple primary cells in the production of electrical energy from chemical energy

Water as a unique chemical

Water is abundant in Earth's environment. Its unique properties allow water to support life in many different ways.

WATER ON EARTH

Water is one of the only substances on Earth that exists at Earth's surface in all three states: gas, liquid and solid. It is estimated that the total volume of water on Earth is approximately 1 247 512 100 km^3. The distribution of this water is shown in Table 2.1.1.

Of all of the water on Earth:

- 97.5% is salt water
- 2.5% is fresh water.

A large proportion of the fresh water is found in ice caps, glaciers or soil. This means the proportion of water available for drinking is less than 1%.

Table 2.1.1 Distribution of water on Earth

Location of water	State	Volume (km^3)
oceans	liquid	1 300 000 000
ice caps and glaciers	solid	24 000 000
groundwater	liquid	23 000 000
ground ice and permafrost	solid	300 000
lakes	liquid	180 7000
soil moisture	liquid	17 000
atmosphere as water vapour	gas	13 000
rivers	liquid	2100

STRUCTURE AND BONDING

Water has several anomalous properties, including:

- relatively high melting and boiling points compared with other group 16 hydrides
- a liquid state that has a higher density than the solid state
- relatively high **specific heat capacity**
- relatively high **latent heat values**.

These anomalous properties can be explained by the structure and bonding in water (Figure 2.1.1).

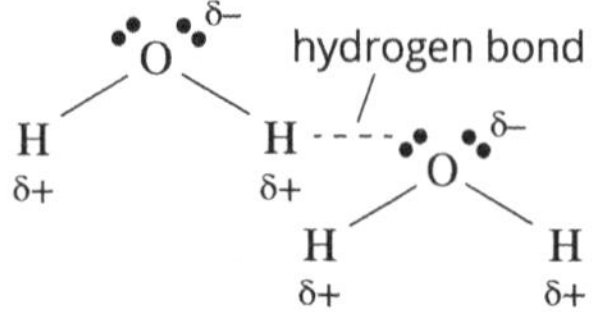

Figure 2.1.1 The structure of water molecules and the bonding between them

The following features are observed.

- Each water molecule contains two hydrogen atoms covalently bonded to a single oxygen atom.
- The oxygen atom in each water molecule has two non-bonding electron pairs.
- The covalent bonds within the molecule are polar; the oxygen atom has a higher electronegativity than the hydrogen atoms. Hence, the oxygen has a greater share of the shared electron pair and carries a negative partial charge. Each hydrogen atom carries a positive partial charge.
- The intermolecular forces between water molecules are hydrogen bonds between the partial positive charge on a hydrogen atom on one molecule and a non-bonding pair of electrons on the oxygen atom of another molecule.

COMPARISON OF BOILING POINT TO GROUP 16 HYDRIDES

Water is an exception to the general trend in melting and boiling points of the **group 16 hydrides** (Table 2.1.2). Generally, the melting and boiling points of the hydrides increase going down the group. This is due to the strength of the intermolecular dispersion forces, which increase as the masses of the molecules increase. Despite being the smallest hydride, water has significantly higher melting and boiling points. This is because each water molecule has the potential to form four hydrogen bonds at a time. A large amount of energy is required to disrupt them.

Table 2.1.2 Melting and boiling points of the group 16 hydrides

Hydride	Melting point (°C)	Boiling point (°C)
H_2O	0	100
H_2S	–82	-60.7
H_2Se	–66	-41.5
H_2Te	–49	-2.2
H_2Po	–35	36.1

DENSITY

The **density** of solid ice is 0.917 g mL^{-3}, whereas the density of liquid water is 0.997 g mL^{-3} at 25°C. The lower density of solid ice means that ice floats in water. As liquid water cools, the water molecules move more slowly. Upon approaching the freezing temperature of water, the molecules arrange so that each water molecule forms four hydrogen bonds to four neighbouring water molecules. This is a very open arrangement of molecules, meaning the water molecules are more widely spaced than in liquid water. In the liquid state, the water molecules are constantly moving and form two or three hydrogen bonds with neighbouring water molecules at any one time.

 ISBN 978 0 6557 0015 9

KEY KNOWLEDGE

HEAT CAPACITY

The heat capacity of a substance is a measure of how much energy a substance absorbs as its temperature increases.

The specific heat capacity of a substance measures the amount of energy (in joules) needed to increase the temperature of a certain amount (usually 1 gram) of that substance by 1°C.

Specific heat capacity is given the symbol c and is expressed in joules per grams per degrees Celsius, i.e. $J\,g^{-1}\,°C^{-1}$.

The specific heat capacity of water is a relatively high $4.18\,J\,g^{-1}\,°C^{-1}$. This is due to the strength of the hydrogen bonding between molecules.

A useful formula links the amount of heat energy added to a substance (q), the mass of the substance (m), the resulting temperature change (ΔT) and the specific heat capacity of the substance (c).

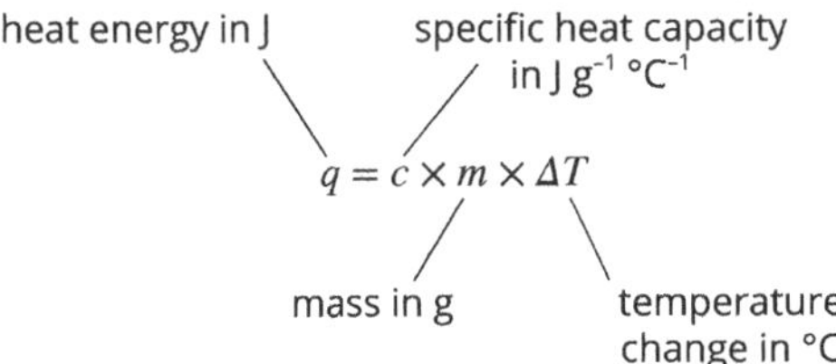

LATENT HEAT

Latent heat is the energy absorbed or released by a substance as it undergoes a change of state.

Latent heat values are a measure of the quantity of heat energy required to melt or boil a given amount of solid or liquid at its melting or boiling temperature. Latent heat values are given the symbol L and can have the units kilojoules per kilogram ($kJ\,kg^{-1}$) or kilojoules per mole ($kJ\,mol^{-1}$) (Table 2.1.3).

Table 2.1.3 Latent heat values for water

Type of latent heat	Definition	L(water) ($kJ\,mol^{-1}$)
latent heat of fusion	the heat needed to change a specified quantity of the substance from a solid to a liquid at its melting point	6.0
latent heat of vaporisation	the heat needed to change a specified quantity of the substance from a liquid to a gas at its boiling point	44.0

When a change of state occurs, a useful formula links the amount of heat energy (q), the amount of substance (n) and the latent heat value of the substance (L).

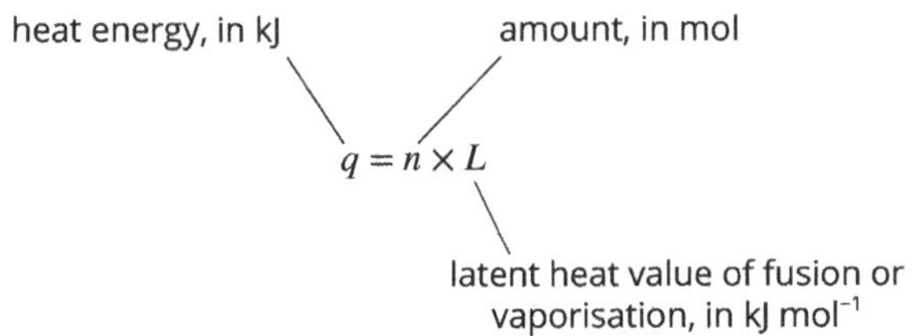

Water's high heat capacity and latent heat of vaporisation has particular significance for aquatic organisms:

- oceans are resistant to sudden temperature changes, which means aquatic organisms do not have to survive sudden changes in surrounding water temperature
- water supplies, including oceans, do not evaporate too quickly.

- **You will now be able to complete Worksheets 23–26 and conduct Practical activity 15.**

Acid–base (proton transfer) reactions

Acids and bases are groups of solutions that have characteristic properties (Table 2.1.4).

Table 2.1.4 Properties of acids and bases

Properties of acids	Properties of bases
taste sour	taste bitter
turn litmus indicator red	turn litmus indicator blue
can be strong or weak	can be strong or weak
react with bases	react with acids
can be corrosive	can be corrosive
solutions have a relatively low pH	solutions have a relatively high pH
solutions conduct an electric current	solutions conduct an electric current

ACIDS

According to the **Brønsted–Lowry theory**, an acid is a proton donor. When an acid molecule is added to water, it ionises by donating a proton, in the form of a H^+ ion, to a water molecule, producing a **hydronium ion, H_3O^+** (Figure 2.1.2).

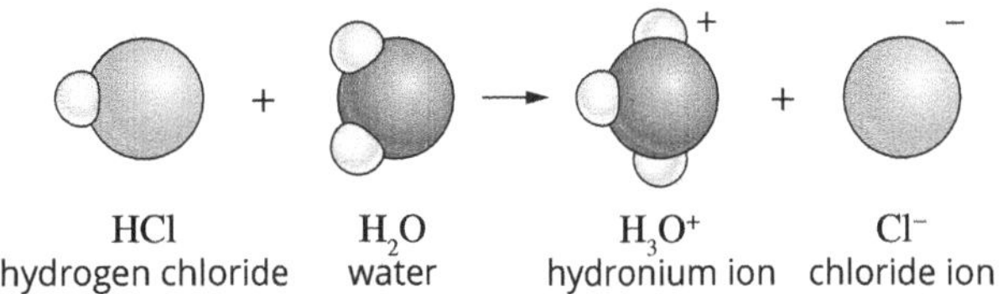

Figure 2.1.2 The reaction between HCl and water: hydrogen chloride is acting as an acid and donating a proton.

The reaction between an acid and water can also be represented in an ionic equation in which spectator ions are omitted. Spectator ions are ions that are in the aqueous state as both reactants and products, so do not take part in the actual reaction.

The ionic equation for the reaction between HCl and water is shown below. $Cl^-(aq)$ is a spectator ion and has been omitted.

$$H^+(aq) + H_2O(l) \rightarrow H_3O^+(aq)$$

KEY KNOWLEDGE

Acids can be strong or weak (Table 2.1.5).

Table 2.1.5 Strong and weak acids

Type of acid	Definition	Example
strong acid	readily donates protons and completely ionises in water	hydrochloric acid, HCl
weak acid	does not readily donate protons and incompletely ionises in water	ethanoic acid, CH_3COOH

Acids and bases can both be strong or weak and dilute or concentrated. The strength of an acid or a base is determined by its ability to ionise in solution, whereas the concentration of an acid or a base refers to the amount present in a given amount of solution (Figures 2.1.3 and 2.1.4).

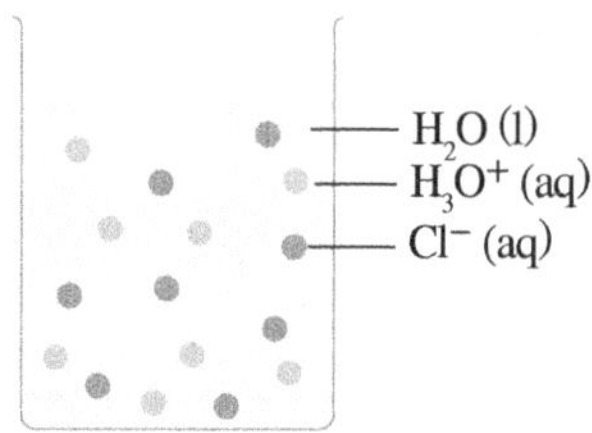

Figure 2.1.3 HCl ionises completely in water. HCl is a strong acid.

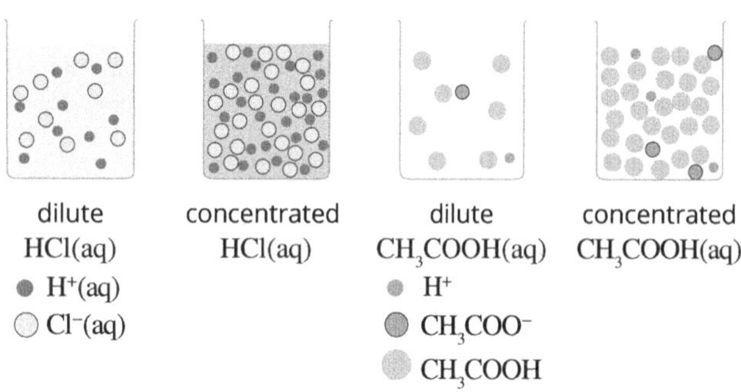

Figure 2.1.4 The concentration of ions in an acid solution depends on both the concentration and strength of the acid.

Different types of acids can donate different numbers of protons (Table 2.1.6). Some acids can donate more than one proton and are referred to as **polyprotic**. However, having more than one hydrogen atom does not automatically make an acid polyprotic; for example, methanoic acid (HCOOH) is a **monoprotic** acid because only one of its hydrogen atoms can be donated to a base.

Table 2.1.6 Monoprotic and polyprotic acids

Type of acid	No. protons it can donate	Example
monoprotic acid	1	nitric acid, HNO_3 ethanoic acid, CH_3COOH
diprotic acid	2	sulfuric acid, H_2SO_4
triprotic acid	3	phosphoric acid, H_3PO_4

BASES

According to the Brønsted–Lowry theory, a base is a proton acceptor. When a base molecule is added to water, it accepts a proton from a water molecule, producing a **hydroxide** ion (OH^-) (Figure 2.1.5).

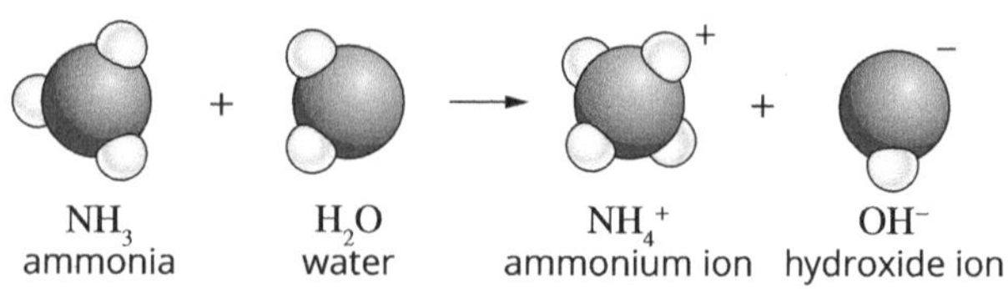

Figure 2.1.5 The reaction between ammonia and water: ammonia acts as a base and accepts a proton.

Bases can be strong or weak (Table 2.1.7 and Figure 2.1.6).

Table 2.1.7 Strong and weak bases

Type of base	Definition	Example
strong base	readily accepts protons	O^{2-}
weak base	does not readily accept protons	NH_3

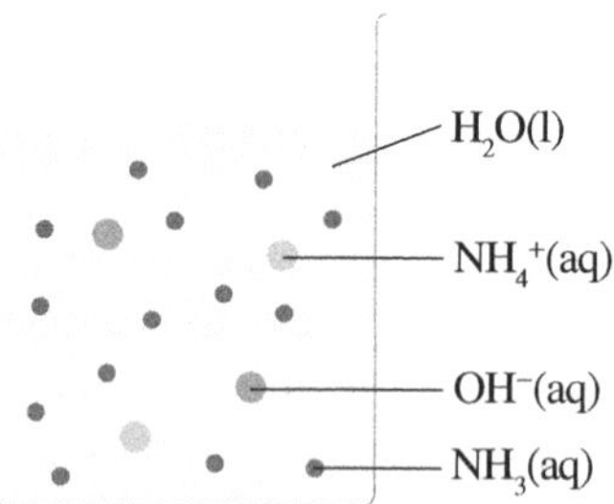

Figure 2.1.6 Only a small proportion of the NH_3 molecules are ionised at any given time. NH_3 is a weak base.

An **amphiprotic** substance can act as either an acid or a base, i.e. it can donate or accept protons. Water is an example of an amphiprotic substance.

- **You will now be able to complete Worksheet 27 and conduct Practical activity 16.**

NEUTRALISATION REACTIONS

A neutralisation reaction is an acid–base reaction in which proton transfer occurs. An acid–base reaction always produces a salt.

For example:

$$\underset{\text{acid}}{HCl(aq)} + \underset{\text{base}}{NaOH(aq)} \rightarrow \underset{\text{salt}}{NaCl(aq)} + \underset{\text{water}}{H_2O(l)}$$

neutral products

This can also be represented in an ionic equation:

$$H^+(aq) + OH^-(aq) \rightarrow H_2O(l)$$

Acids undergo neutalisation reactions with many substances to produce salts. Table 2.1.8 lists several common types of reactions of acids.

 ISBN 978 0 6557 0015 9

KEY KNOWLEDGE

Table 2.1.8 Common reactions of acids

Reactants	Products
acid + metal	salt + hydrogen gas
acid + metal hydroxide	salt + water
acid + metal carbonate	salt + water + carbon dioxide gas

Antacids are a medication taken to treat heartburn (indigestion), which is caused by stomach acids entering the lower end of the oesophagus. Antacids react with stomach acid in a neutralisation reaction to produce salt, thereby easing the symptoms. For example, calcium carbonate is a common ingredient in antacids that reacts with hydrochloric acid in the stomach according to the equation:

$$2HCl(aq) + CaCO_3(aq) \longrightarrow CaCl_2(aq) + H_2O(l) + CO_2(g)$$

Ionic equation:

$$2H^+ + CO_3^{2-}(aq) \longrightarrow H_2O(l) + CO_2(g)$$

pH SCALE

The **pH scale** (Figure 2.1.7) indicates the acidity of a solution. In general, at 25°C:

- pH < 7 is an acidic solution
- pH = 7 is a neutral solution
- pH > 7 is a basic solution.

The more acidic a solution, the lower the pH; the more basic a solution, the higher the pH.

pH values are a convenient way of specifying the acidity of a solution because they are dependent on the concentration of hydronium ions present:

$$pH = -\log_{10}[H_3O^+]$$

When strong acids are added to water, they completely ionise to produce hydronium ions. Therefore, a direct connection exists between the amount of acid molecules added to water, the number of protons each molecule is able to donate and the pH of the solution.

Pure water undergoes a small amount of **self-ionisation**; that is, it reacts with itself to form an equal number of hydronium and hydroxide ions according to the equation:

$$H_2O(l) + H_2O(l) \longrightarrow H_3O^+(aq) + OH^-(aq)$$

In any solution, the ionic product of water, K_w, at 25°C is $[H_3O^+][OH^-] = 10^{-14} M^2$

When bases are added to water, they produce hydroxide ions. The concentration of hydronium ions is directly affected and can be calculated at 25°C by using the ionic product of water:

$$[H_3O^+] = \frac{10^{-14}}{[OH^-]}$$

Therefore, there is a direct relationship between the number of base molecules added to water, the number of hydroxide ions each molecule produces and the pH of the solution.

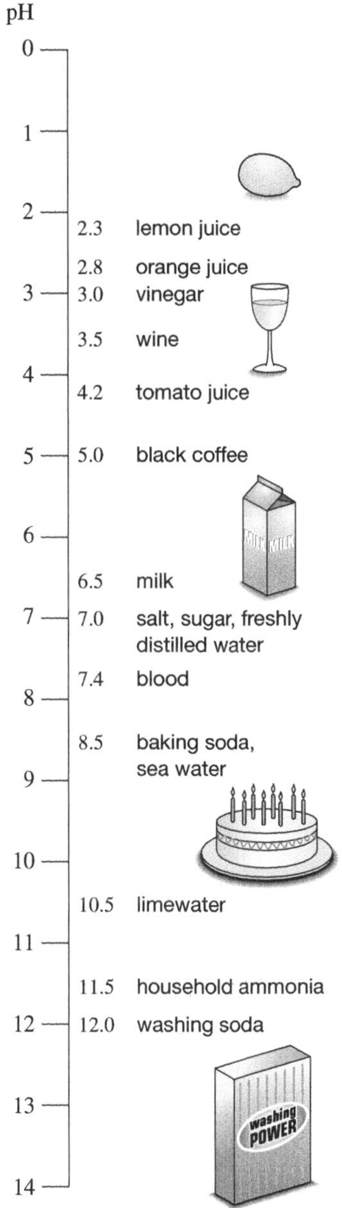

Figure 2.1.7 The pH scale and the pH of some common household substances

METHODS TO MEASURE pH

Indicators are weak acids that change colour depending on the pH. They are used to determine the acidity or basicity of a solution.

- **Natural indicators** are plant extracts that vary in colour depending on the pH of the solution. An example is red cabbage.
- **Commercial indicators** are prepared for use in laboratories and industry. They tend to show a distinct colour change at a certain pH. An example is phenolphthalein, which is colourless at a pH < 8.3 and pink at pH > 10.0. Universal indicator is a mixture of several indicators and displays a range of colours from pH 0 to 14.
- A **pH meter** is a piece of equipment that provides a digital read-out of the pH of a solution and is not dependent on colour change.

KEY KNOWLEDGE

Different methods of measuring pH have different levels of accuracy and precision (Table 2.1.9). The accuracy of a pH value is an indication of how close the measured value is to the true value. The precision of a pH value is an indication of the number of significant figures in the measured value, as well as how well repeated measurements produce the same number.

Table 2.1.9 Accuracy and precision in measurement of pH

Method of determining pH	Accuracy	Precision
natural indicator such as red cabbage juice	sample value can be determined to within 2 pH units of the true value	generally reproducible to within 2 pH units
commercial indicator such as phenolphthalein	sample value can be determined to within 1 pH unit of the true value	generally reproducible to within 1 pH unit
pH meter	sample value can be determined to within 0.01 pH units of the true value.	generally reproducible to within 0.01 pH units of the true value.

APPLICATIONS OF ACID-BASE REACTIONS

In addition to antacids, there are many examples of acid–base reactions in society. This includes in cooking, in industry and in the environment such as acid rain and ocean acidification.

Ocean acidification

Carbon dioxide dissolves in ocean water, forming a weak acid. The amount of carbon dioxide that dissolves increases with increasing atmospheric carbon dioxide, which can be attributed to the enhanced greenhouse effect. Increasing acidity of the oceans affects the shell growth in marine invertebrates because their shells are made from calcium carbonate. Some of the additional hydronium ions react with carbonate ions, CO_3^{2-}, in the following equation:

$$H_3O^+(aq) + CO_3^{2-}(aq) \longrightarrow HCO_3^-(aq) + H_2O(l)$$

This reduces the concentration of carbonate ions in the water, making it more difficult for marine invertebrates to build or maintain their protective structures.

- **You will now be able to complete Worksheets 28 and 29 and conduct Practical activities 17 and 18.**

Redox (electron transfer) reactions

Redox reactions involve the transfer of electrons.

Redox reactions occur in two parts:

- **oxidation**—the loss of electrons
- **reduction**—the gain of electrons.

Oxidation and reduction occur simultaneously. The electrons lost by the species undergoing oxidation are the electrons gained by the species undergoing reduction. The species undergoing oxidation is called the **reducing agent**. The species undergoing reduction is called the **oxidising agent** (Figure 2.1.8).

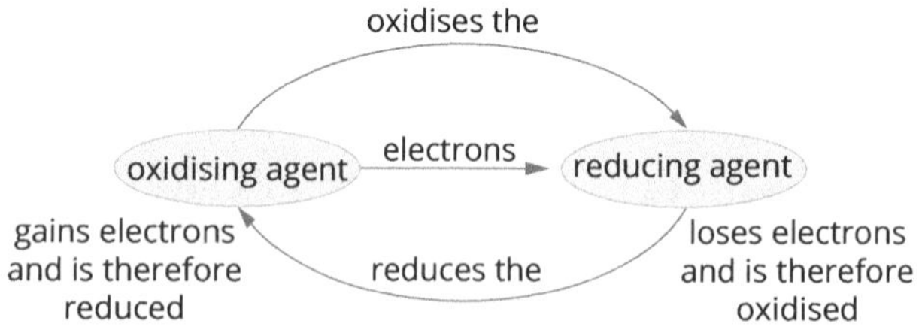

Figure 2.1.8 The relationship between oxidising and reducing agents

WRITING REDOX REACTIONS

Half-equations represent the separate processes of oxidation and reduction. Half-equations are balanced in terms of:

- elements: the number and type of atoms on the reactant side of the equation must equal the number and type on the product side
- charge: electrons are represented by the symbol e^- and are added to balance the charge on either the reactant or product side as needed.

States are also included. For example, in the reaction between copper metal and silver chloride, copper is oxidised to copper ions and silver ions are reduced.

Oxidation: $Cu(s) \longrightarrow Cu^{2+}(aq) + 2e^-$ ↑ reducing agent	Electrons are lost, so appear as a product. They balance the charge on the copper ion. Two electrons are lost for each copper atom oxidised.
Reduction: $Ag^+(aq) + e^- \longrightarrow Ag(s)$ ↑ oxidising agent	An electron is gained, so it appears as a reactant. It balances the charge on the silver ion. One electron is gained to reduce each silver ion.

 ISBN 978 0 6557 0015 9

KEY KNOWLEDGE

An overall equation is written as a combination of half-equations. The overall equation is balanced so that the number of electrons lost in the oxidation reaction is equal to the number of electrons gained in the reduction reaction. In the example above, two silver ions will be reduced for every copper atom oxidised. Electrons do not appear in the overall equation:

$$Cu(s) + 2Ag^+(aq) \longrightarrow Cu^{2+}(aq) + 2Ag(s)$$

More complex half-equations occurring in acidic solutions are balanced according to the rules in Table 2.1.10.

Table 2.1.10 Rules for balancing complex redox half-equations in acidic solutions

Rule	Example: reduction of $Cr_2O_7^{2-}(aq)$ to $Cr^{3+}(aq)$
1 Balance all elements except hydrogen and oxygen.	$Cr_2O_7^{2-}(aq) \longrightarrow 2Cr^{3+}(aq)$
2 Balance oxygen atoms by adding H_2O molecules.	$Cr_2O_7^{2-}(aq) \longrightarrow 2Cr^{3+}(aq) + 7H_2O(l)$
3 Balance hydrogen atoms using H^+ ions.	$Cr_2O_7^{2-}(aq) + 14H^+(aq) \longrightarrow 2Cr^{3+}(aq) + 7H_2O(l)$
4 Balance charge, using electrons, and add states.	$Cr_2O_7^{2-}(aq) + 14H^+(aq) + 6e^- \longrightarrow 2Cr^{3+}(aq) + 7H_2O(l)$

REACTIVITY SERIES OF METALS

Metals always lose electrons to form ions; that is, they always undergo oxidation, so they act as reducing agents. A **reactivity series** of metals can be presented as an ordered listing of reduction half-equations (Figure 2.1.9), which can be used to predict redox reactions. The metal atoms are on the right-hand side. The reactivity, or reducing strength, of the metals increases going down the series.

The order of these half-equations is determined at standard laboratory conditions (25°C, 1 M concentrations), so predictions under other conditions may not be reliable.

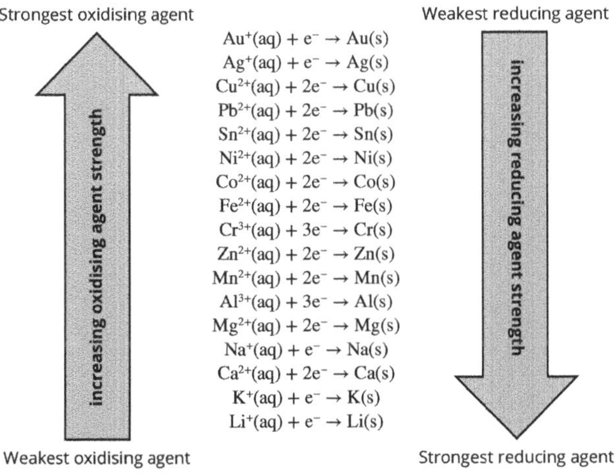

Figure 2.1.9 The reactivity series of metals

METAL DISPLACEMENT REACTIONS

The likelihood of a **metal displacement** reaction occurring between a metal and another metal ion can be predicted by comparing the relative positions of the metal ion (potential oxidising agent) and the metal (potential reducing agent) in the reactivity series. A more reactive metal will be oxidised by, and donate its electrons to, the cation of a less reactive metal. A reaction will only occur if the reducing agent is lower in the series than the oxidising agent (Figure 2.1.10).

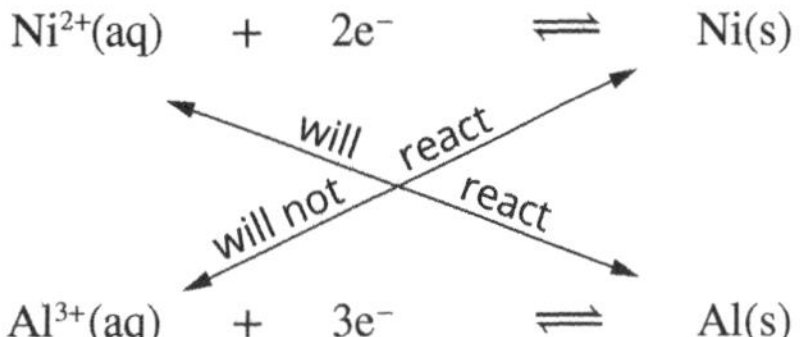

Figure 2.1.10 Al metal will react with Ni^{2+} ions; however, Ni metal will not react with Al^{3+} ions.

Balanced full and ionic equations can then be written for predicted reactions by balancing the electrons lost and gained. For example, in a reaction between aluminium metal and nickel(II) nitrate:

ionic equation:

$$2Al(s) + 3Ni^{2+}(aq) \longrightarrow 3Ni(s) + 2Al^{3+}(aq)$$

full equation:

$$2Al(s) + 3Ni(NO_3)_2(aq) \longrightarrow 3Ni(s) + 2Al(NO_3)_3(aq)$$

APPLICATIONS OF REDOX REACTIONS

Simple primary cells

Primary cells are simple galvanic cells commonly used as non-rechargeable batteries in many portable electronic devices. Galvanic cells are electrochemical cells that convert chemical energy to electrical energy. They involve spontaneous redox reactions. The two half-reactions occur in two separate half-cells. This prevents direct contact between the oxidising agent and reducing agent, so electrons can only be transferred by travelling through an external circuit, between the negative and positive electrodes. The flow of electrons provides electrical energy.

A simple galvanic cell has the features labelled in Figure 2.1.11 on page 116.

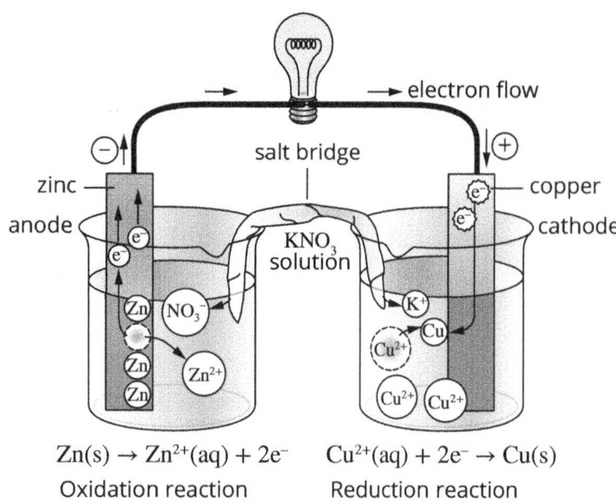

Figure 2.1.11 The general features of a galvanic cell

In the cell in Figure 2.1.11, the zinc metal electrode is oxidised to zinc ions:

$$Zn(s) \rightarrow Zn^{2+}(aq) + 2e^-$$

The electrons produced move through the wire (the external circuit) towards the positive electrode, where they reduce copper ions:

$$Cu^{2+}(aq) + 2e^- \rightarrow Cu(s)$$

The number of electrons produced is equal to the number of electrons consumed at the other half-cell. The overall equation is the sum of the two half-equations:

$$Cu^{2+}(aq) + Zn(s) \rightarrow Cu(s) + Zn^{2+}(aq)$$

Zn is oxidised and acts as the reducing agent. Cu^{2+} is reduced and acts as the oxidising agent. The salt bridge (often a piece of filter paper soaked in a solution of a soluble ionic compound) balances the charges formed and consumed in each half-cell. At the negative electrode, the positive Zn^{2+} ions being formed are balanced by negative ions moving from the salt bridge. At the positive electrode, the Cu^{2+} ions being consumed are balanced by positive ions from the salt bridge.

By definition, the electrode where oxidation occurs is the **anode** and the electrode where reduction occurs is the **cathode**. In galvanic cells, the anode is negative and the cathode is positive. Each half-cell contains a conjugate redox pair. In the cell in Figure 2.1.11, the pairs can be written as $Zn^{2+}(aq)/ Zn(s)$ and $Cu^{2+}(aq)/ Cu(s)$.

- **You will now be able to complete Worksheets 30–32 and conduct Practical activities 19 and 20.**

 ISBN 978 0 6557 0015 9

Knowledge review—identifying and naming types of substances, balancing chemical equations

1 A chemical reaction involves the rearrangement of particles in a substance to make up a new substance. Bonds must be broken and re-formed. Identify the type(s) of bonding present in each of the substances in the table.

Substance	Type(s) of bonding
magnesium	
aluminium nitride	
hydrogen chloride	
oxygen	
iron	
iron oxide	
carbon dioxide	
water	
diamond	
sodium hydroxide	

2 When a chemical reaction occurs, mass is conserved.

Which subatomic particles make a significant contribution to the mass of an atom?

3 In the following reaction, 5.0 g of magnesium reacts with oxygen to produce magnesium oxide:

$$2Mg(s) + O_2(g) \rightarrow 2MgO(s)$$

Calculate the percentage by mass of magnesium in magnesium oxide and predict the mass of magnesium oxide formed.

4 Balanced chemical equations demonstrate that the type and number of atoms in the reactants are also present in the products. Balance each of the following equations.

$HCl(aq) + MgCO_3(aq) \rightarrow MgCl_2(aq) + H_2O(l) + CO_2(g)$

$C_4H_{10}(g) + O_2(g) \rightarrow CO_2(g) + H_2O(g)$

$Al(s) + O_2(g) \rightarrow Al_2O_3(s)$

ISBN 978 0 6557 0015 9

Structure and properties of water

1 Draw the structures of five water molecules as they might appear in a sample of liquid water. Include the non-bonding electron pairs.

2 a What type of bonding holds the oxygen and hydrogen atoms together within the molecule? Label one of these bonds on your diagram in Question 1.

b Explain why water is a polar molecule.

c The attraction between different water molecules is hydrogen bonding. Label a hydrogen bond on your diagram in Question 1.

d Describe, specifically, the parts of the molecules between which hydrogen bonding occurs.

e List three other substances that have hydrogen bonds between molecules.

f Why is hydrogen bonding so strong compared with other types of dipole–dipole bonding?

3 Consider the photograph, and identify the unique property of water being demonstrated. Explain why water demonstrates this unique property.

 ISBN 978 0 6557 0015 9

WORKSHEET 25

Calculations using specific heat capacity

1 Choose from the terms below to complete the summary statements outlining the key points about specific heat capacity.

energy	1°C	c	$J\,g^{-1}\,°C^{-1}$	specific heat capacity
change in temperature	mass	$4.18\,J\,g^{-1}\,°C^{-1}$	q	$q = c \times m \times \Delta T$

a The specific heat capacity of a substance represents the amount of ______________ needed to increase the temperature of a specified amount by ______________.

b Specific heat capacity is given the symbol ______________ and the unit ______________. The specific heat capacity of water is ______________.

c Heat energy is given the symbol ______________. The heat energy, in joules, required to produce a particular temperature increase for a given mass of substance can be calculated using the formula:

____

where c is the ______________, m is the ______________ and ΔT is the ______________.

2 Complete the calculations below using the specific heat capacity of water.

a Calculate the heat energy, in kJ, needed to increase the temperature of 240 g of water from 18.0°C to 100°C.

b Calculate the heat energy, in kJ, needed to increase the temperature of 3.0 L of water from 25.0°C to 100°C.

3 Calculate the increase in temperature that will occur when 5.0 kJ of energy is added to 200 mL of water.

4 Calculate the mass of water that can be heated from 10°C to 35°C by the addition of 1000 J of energy.

Calculations using latent heat

1 Read the definitions listed in the boxes on the right of the page. Choose the correct term from the list below to match each definition. Write each term in the box corresponding to the definition.

latent heat of vaporisation	latent heat of fusion	latent heat
44.0 kJ mol^{-1}	6.0 kJ mol^{-1}	

Term	Definition
	the heat needed to change one mole of a substance from a solid to a liquid at its melting point
	the heat needed to change one mole of a substance from a liquid to a gas at its boiling point
	the energy absorbed by a substance as it changes state from a solid to a liquid or a liquid to a gas
	the latent heat of vaporisation of water
	the latent heat of fusion of water

2 **a** The amount of heat energy required to change the state of a specific mass of water is calculated using the formula $q = n \times L$, where q is the ____________ in kilojoules, kJ, n is the __________, in mol, of water and L is the __________, measured in kJ mol^{-1}.

b Calculate the heat energy, in kJ, required to evaporate 50.0 g of water at 100°C.

c Calculate the heat energy, in kJ, required to melt 1.0 kg of ice at 0°C.

d Calculate the mass of water, in g, that can be evaporated at 100°C by the addition of 3500 kJ.

 ISBN 978 0 6557 0015 9

Concentration and strength—picturing acids and bases

Complete the table. The first row has been done for you.

Solution	Weak or strong acid or base?	Particles present in solution	Sketch showing relative proportions of particles
concentrated hydrochloric acid	strong acid	H_2O molecules Cl^- ions H_3O^+ ions	● $H_3O^+(aq)$ ○ $Cl^-(aq)$ concentrated $HCl(aq)$
dilute nitric acid			
concentrated ethanoic acid			
dilute sodium hydroxide			
concentrated ammonia			

ISBN 978 0 6557 0015 9

Predicting products of acid reactions

Acids demonstrate similar chemical behaviour. They react with many metals and metal compounds in common ways.

1 Complete the following table by writing the names of the predicted products of each reaction.

Reaction number	Reactants	Products
1	hydrochloric acid and zinc metal	
2	nitric acid and aqueous potassium hydroxide	
3	sulfuric acid and sodium carbonate solution	
4	hydrochloric acid and solid magnesium oxide	

2 Write balanced full and ionic chemical equations for the reactions in the table.

Reaction 1

Full: ______________________________

Ionic: ______________________________

Reaction 2

Full: ______________________________

Ionic: ______________________________

Reaction 3

Full: ______________________________

Ionic: ______________________________

Reaction 4

Full: ______________________________

Ionic: ______________________________

3 In which of the reactions would you expect to observe bubbles being formed? Explain your answer.

____

____

____

4 In which of the reactions would you expect to observe a change in pH? Explain your answer.

____

____

____

 ISBN 978 0 6557 0015 9

Calculating pH

The table summarises some steps you can follow when calculating the pH of different types of strong acids and strong bases from their concentrations.

Solution	Step 1	Step 2	Step 3
monoprotic acid	$[H_3O^+] = [acid]$	$pH = -\log_{10}[H_3O^+]$	
diprotic acid	$[H_3O^+] = 2 \times [acid]$	$pH = -\log_{10}[H_3O^+]$	
base that ionises to produce one OH^- ion	$[OH^-] = [base]$	$[H_3O^+] = \frac{10^{-14}}{[OH^-]}$	$pH = -\log_{10}[H_3O^+]$
base that ionises to produce two OH^- ions	$[OH^-] = 2 \times [base]$	$[H_3O^+] = \frac{10^{-14}}{[OH^-]}$	$pH = -\log_{10}[H_3O^+]$

1 Determine the pH of each solution by following these steps:

i Identify whether the substance is an acid or a base.

ii Determine whether the substance donates 1, 2 or 3 protons (if an acid) or ionises to produce 1, 2 or 3 hydroxide ions (if a base).

iii Follow the appropriate steps as outlined in the table above.

The first one has been done for you.

a 0.10 M HCl

i acid

ii monoprotic

iii $[H_3O^+] = [acid]$

$= 0.10\,M$

$pH = -\log_{10}[H_3O^+]$

$= -\log_{10}(0.10)$

$= 1.00$

b 0.20 M HCl

i ______________

ii ______________

iii ______________

c 0.10 M NaOH

i ______________

ii ______________

iii ______________

d 0.010 M $Ca(OH)_2$

i ______________

ii ______________

iii ______________

2 Calculate the concentration, in M, of each solution by following these steps:

i Identify whether the substance is an acid or a base.

ii Determine whether the substance donates 1, 2 or 3 protons (if an acid) or ionises to produce 1, 2 or 3 hydroxide ions (if a base).

iii Follow, in reverse, the appropriate steps as outlined in the table.

The first one has been done for you.

ISBN 978 0 6557 0015 9

a HCl solution with a pH of 2.00

i acid

ii monoprotic

iii $pH = -\log_{10}[H_3O^+]$

$[H_3O^+] = 10^{-pH}$

$= 10^{-2}$ M

$[acid] = [H_3O^+]$

$= 0.010$ M

b HNO_3 solution with a pH of 1.00

i ______________

ii ______________

iii ______________

c NaOH solution with a pH of 12.00

i ______________

ii ______________

iii ______________

d $Ca(OH)_2$ solution with a pH of 11.00

i ______________

ii ______________

iii ______________

3 Write a summary of at least one sentence about the difference between acids and bases.

4 Write a summary of at least one sentence about the differences between monoprotic and diprotic acids.

5 You have a solution of unknown pH.

a Describe how you might obtain a pH measurement with low precision.

b Describe how you would obtain a pH measurement with the highest accuracy and precision.

 ISBN 978 0 6557 0015 9

WORKSHEET 30

Redox reactions and reactivity of metals

A student places a piece of copper metal in a beaker of silver nitrate solution. The solution slowly turns a blue colour.

This solution has turned blue.

1 What ion is causing the blue colour in the solution?

____

2 What do you expect the student to observe on the surface of the copper metal?

____

3 Atoms and their ions are species with very different properties. Complete the electron-shell diagrams by adding electrons, and write each of the following words below the correct diagram to show some different properties of the copper atom and the copper(II) ion.

metal	ion	insoluble	soluble	blue	lustrous

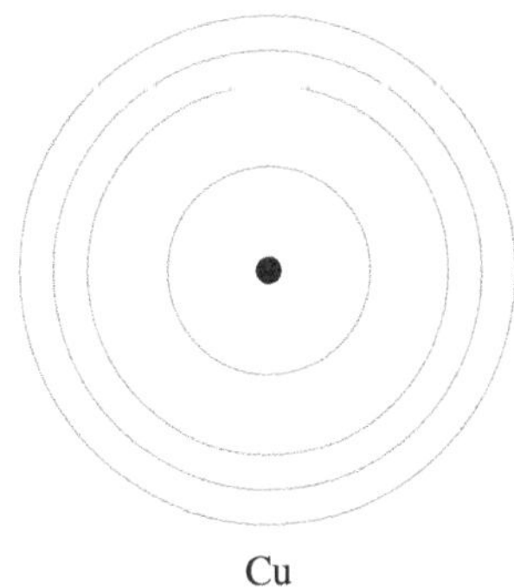

Cu

Properties ______________________________

____

____

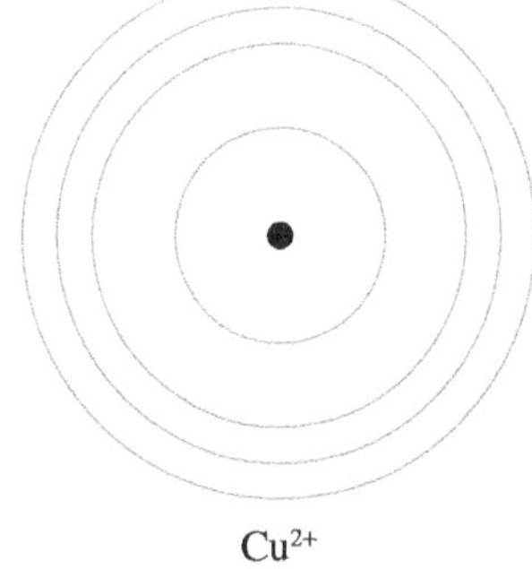

Cu^{2+}

Properties ______________________________

____

____

ISBN 978 0 6557 0015 9

4 Complete the half-equations for the oxidation of copper metal and reduction of silver ions by adding one or more electrons where required.

$Cu(s) \rightarrow Cu^{2+}(aq)$

$Ag^{+}(aq) \rightarrow Ag(s)$

5 Write a balanced chemical equation for the overall reaction.

6 Identify the oxidising agent and reduction agent in the reaction in Question 4.

A reactivity series of metals is an ordered listing of reduction half-equations, which is also called an electrochemical series. It can be used to predict redox reactions.

$$\begin{array}{l} Au^{+}(aq) + e^{-} \rightarrow Au(s) \\ Ag^{+}(aq) + e^{-} \rightarrow Ag(s) \\ Cu^{2+}(aq) + 2e^{-} \rightarrow Cu(s) \\ Pb^{2+}(aq) + 2e^{-} \rightarrow Pb(s) \\ Sn^{2+}(aq) + 2e^{-} \rightarrow Sn(s) \\ Ni^{2+}(aq) + 2e^{-} \rightarrow Ni(s) \\ Co^{2+}(aq) + 2e^{-} \rightarrow Co(s) \\ Fe^{2+}(aq) + 2e^{-} \rightarrow Fe(s) \\ Cr^{2+}(aq) + 3e^{-} \rightarrow Cr(s) \\ Zn^{2+}(aq) + 2e^{-} \rightarrow Zn(s) \\ Mn^{2+}(aq) + 2e^{-} \rightarrow Mn(s) \\ Al^{3+}(aq) + 3e^{-} \rightarrow Al(s) \\ Mg^{2+}(aq) + 2e^{-} \rightarrow Mg(s) \\ Na^{+}(aq) + e^{-} \rightarrow Na(s) \\ Ca^{2+}(aq) + 2e^{-} \rightarrow Ca(s) \\ K^{+}(aq) + e^{-} \rightarrow K(s) \end{array}$$

increasing reactivity

Use the electrochemical series shown to complete the following:

7 Give the formula of the strongest reducing agent.

8 Give the formula of the weakest reducing agent.

9 Give the formula of the strongest oxidising agent.

10 Give the formula of the weakest oxidising agent.

11 Predict whether the following reactions will occur by placing a tick or cross in the table. For those that will occur, write full and ionic equations.

Substances	Prediction of whether a reaction will occur (✓ or ✗)	Equations
silver nitrate and tin		
aluminium and nickel sulfate		
lead nitrate and calcium sulfate		
zinc and copper(II) nitrate		
potassium and sodium		

 ISBN 978 0 6557 0015 9

WORKSHEET 31

Literacy review—matching redox key terms

1 Read the definitions listed in the boxes on the right of the page. Choose the correct term from the list below to match each definition. Write each term in the box corresponding to its definition.

reducing agent	oxidation	ion	oxidising agent	reduction
reactivity series of metals	electron	OILRIG	redox reaction	half-equation

Term	Definition
	substance that causes another substance to undergo an oxidation reaction and is itself reduced
	substance that causes another substance to undergo a reduction reaction and is itself oxidised
	negatively charged subatomic particle that moves around the nucleus of an atom (in a reaction, it is given the symbol e^-)
	reaction in which a substance loses one or more electrons
	reaction in which a substance gains one or more electrons
	reaction in which electrons are transferred
	atom (or group of atoms) that carries a charge
	a list of metals ordered according to their reducing strength
	equation showing either oxidation or reduction; electrons are shown in the equation
	useful way to remember that Oxidation Is Loss and Reduction Is Gain

2 Write a paragraph in your own words that includes all of the terms listed and defined in Question 1 and explains the concepts of redox chemistry.

ISBN 978 0 6557 0015 9

WORKSHEET 32

Reflection—How do chemicals interact with water?

The following table lists the key knowledge covered in this area of study.

1 Reflect on how well you understand the concepts listed. Rate your learning by shading the circle that corresponds to your current level of understanding for each one.

Key knowledge	Not confident ◄				► Very confident
Water—anomalous properties, specific heat capacity and latent heat	○	○	○	○	○
Definitions of acids and bases, polyprotic and amphiprotic species, neutralisation reactions	○	○	○	○	○
pH scale—calculating pH, measuring pH using indicators	○	○	○	○	○
Applications of acid–base reactions in society	○	○	○	○	○
Redox reactions—definitions of oxidising and reducing agents, writing balanced half and overall equations	○	○	○	○	○
Reactivity series of metals—predicting metal displacement reactions and writing equations	○	○	○	○	○
Applications of redox reactions in society	○	○	○	○	○

2 Consider the points you have shaded from 'Not confident' to 'Very confident'. List specific ideas you can identify that were challenging.

3 Write down two different strategies that you will apply to help further your understanding of these ideas.

 ISBN 978 0 6557 0015 9

PRACTICAL ACTIVITY 15

Controlled experiment

Density of water and ice

SUGGESTED DURATION

- 45 minutes over 2 days

INTRODUCTION

Water has unique physical and chemical properties that make it essential for life. The expansion of water on freezing causes icebergs to float and water pipes to burst in cold climates.

AIM

- To determine the density of water at room temperature and the density of ice
- To compare the effect of cooling on the density of water with the effect of cooling on the density of other liquids

MATERIALS

- 120 mL deionised water
- ice cubes
- 15 mL cooking oil
- 120 mL methylated spirits
- 3 × 10 mL glass measuring cylinders
- 2 × 100 mL glass measuring cylinders
- 2 × 250 mL glass beakers
- electronic balance
- freezer
- thermometer (–10°C to 110°C)
- tomato
- lettuce
- safety glasses/goggles
- disposable gloves

PRE-LAB SAFETY INFORMATION

Material used/other risks	Hazard	Control
methylated spirits	• highly flammable liquid and vapour • causes serious eye irritation • may cause damage to organs if ingested	Wear eye and skin protection. Do not use near ignition sources.

Please indicate that you have understood the information in the safety table.

Name (print): ______________________________

I understand the safety information (signature): ______________________________

METHOD

1. Put on your PPE, ensuring your lab coat is buttoned, your gloves are fitted correctly and comfortably, and your safety glasses/goggles are fitted over your eyes.
2. Place one cube of ice into a beaker containing 100 mL of cold water. Place the other cube into a beaker containing 100 mL of methylated spirits.
3. Estimate the percentage volume of ice cube that is protruding above the surface of the water. Perform the same estimation for the methylated spirits.
4. Weigh one of the measuring cylinders. Fill the measuring cylinder to the 9.0 mL mark with deionised water and reweigh it. Measure and record the temperature of a sample of deionised water.
5. Weigh the other two measuring cylinders. Fill one measuring cylinder to the 9.0 mL mark with oil and the other with methylated spirits, and reweigh them. Measure and record the temperature of a sample of cooking oil and a sample of methylated spirits.
6. Place the three cylinders in the freezer section of a refrigerator or in a deep freezer for at least 24 hours.
7. At the same time, put some sliced tomato and some lettuce in the freezer and leave them there for at least 24 hours.
8. Take the measuring cylinders out of the freezer and immediately record the volume occupied by the solid or cooled liquid in each.
9. Remove the tomato and lettuce from the freezer and carefully examine their appearance. Leave them to thaw out and then re-examine them.
10. Spray and wipe down your bench, return used equipment, wash equipment as necessary, and remove PPE as directed by your teacher.

ISBN 978 0 6557 0015 9

PRACTICAL ACTIVITY 15

RESULTS

Record your observations and results in the space below. Draw appropriate tables to record results.

DISCUSSION

1 Given that density is mass/volume, calculate the density, in $g\,mL^{-1}$ of:

a water, oil and methylated spirits at room temperature

b water, oil and methylated spirits after being in the freezer for 24 hours

2 Compare the effect of cooling on the densities of each of the three substances.

3 Explain what happens to the arrangement of water molecules when ice melts.

4 Describe the appearance of the tomato and lettuce before they were frozen, when they were frozen and when they were defrosted. Account for any changes in appearance.

CONCLUSION

ISBN 978 0 6557 0015 9

Investigating acids

SUGGESTED DURATION

- 30 minutes

INTRODUCTION

Acid strength is related to how readily the acid donates a proton. Different acids have different strengths.

Polyprotic acids are able to donate more than one proton. Amphiprotic substances can act as both acids and bases. In an aqueous solution, the molecules of an amphiprotic substance will undergo one of two hydrolysis reactions, one as an acid and one as a base. In most cases, one of these reactions will predominate. If the acid reaction occurs to a greater extent than the base reaction, $[H_3O^+] > [OH^-]$, and the pH of the solution will be less than 7. If the base reaction occurs more than the acid reaction, then the pH of the solution will be greater than 7.

In this experiment, the sodium salt of each amphiprotic anion is used. The $Na^+(aq)$ ion does not hydrolyse to any appreciable extent in water. Therefore, any change of pH can be attributed to the acid–base reactions of the amphiprotic anion in each solution.

AIM

PART A • STRONG, WEAK AND POLYPROTIC ACIDS

To compare the relative strengths of identical concentrations of some naturally occurring and manufactured acids in order to:

- relate the measured pH of acids to the concentration of H_3O^+ ions
- distinguish between some strong and weak acids
- distinguish between monoprotic and polyprotic acids

PART B • AMPHIPROTIC SUBSTANCES

To determine the relative acid and base strengths of some amphiprotic substances

PART C • NEUTRALISATION

To record the pH changes that occur during a neutralisation reaction

PRE-LAB SAFETY INFORMATION		
Material used/other risks	**Hazard**	**Control**
0.1 M hydrochloric acid	• non-hazardous < 3 M	Wear safety glasses/goggles.
0.1 M sulfuric acid	• mild eye irritant	Wear safety glasses/goggles.
0.1 M phosphoric acid	• non-hazardous	
universal indicator	• flammable liquid and vapour • causes eye irritation	Wear safety glasses/goggles.
Please indicate that you have understood the information in the safety table. Name (print): I understand the safety information (signature):		

MATERIALS

- 20 mL of 0.1 M hydrochloric acid, HCl
- 20 mL of 0.1 M ethanoic acid, CH_3COOH
- 20 mL of 0.1 M sulfuric acid, H_2SO_4
- 20 mL of 0.1 M phosphoric acid, H_3PO_4
- dropper bottles of:
  - 0.1 M $NaHCO_3$
  - 0.1 M $NaHSO_4$
  - 0.1 M Na_2HPO_4
  - 0.1 M NaH_2PO_4
  - 0.1 M NaOH
- universal indicator and pH chart
- pH meter or pH probe with data logger
- semi-micro test tubes
- small beakers if pH meter/probe is to be used
- safety glasses/goggles

ISBN 978 0 6557 0015 9

METHOD

Put on your PPE, ensuring that your safety glasses/goggles are fitted over your eyes.

PART A • STRONG, WEAK AND POLYPROTIC SUBSTANCES

1. Add about 20 mL of each acid to a small beaker.
2. Use the pH meter to record the pH of each acid in Table 1.
3. Calculate the concentration of H_3O^+ ions in each of the acids.

RESULTS

Table 1 pH and $[H_3O^+]$ of acids

Acid	pH	$[H_3O^+]$
0.1 M hydrochloric acid		
0.1 M ethanoic acid		
0.1 M sulfuric acid		
0.1 M phosphoric acid		

PART B • AMPHIPROTIC SUBSTANCES

1. Place 20 drops of each of 0.1 M $NaHCO_3$, $NaHSO_4$, Na_2HPO_4 and NaH_2PO_4 into separate semi-micro test tubes. Add one drop of universal indicator to each solution. Record the colour of each solution in Table 2. Alternatively, a pH meter can be used with small beakers.
2. Use the indicator chart to determine the pH of each solution.

RESULTS

Table 2 Colour and pH of amphiprotic substances

Solution	Amphiprotic anion	Colour of universal indicator	pH
0.1 M $NaHCO_3$	HCO_3^-		
0.1 M $NaHSO_4$	HSO_4^-		
0.1 M Na_2HPO_4	HPO_4^{2-}		
0.1 M NaH_2PO_4	$H_2PO_4^-$		

PART C • NEUTRALISATION REACTION

1. Add one drop of universal indicator to a test tube containing 20 drops of 0.1 M HCl. Add 0.1 M NaOH drop by drop and record the colour changes in Table 3

 ISBN 978 0 6557 0015 9

RESULTS

Table 3 Universal indicator colour changes during a neutralisation reaction

Drops of NaOH added	Observations
1	
2	
3	
4	
5	
6	
7	
8	
9	
10	
11	
12	
13	
14	
15	
16	
17	
18	
19	
20	

DISCUSSION

1 Rank the solutions tested in Part A from most acidic to least acidic.

2 a Account for the difference between the measured pH values of ethanoic acid and hydrochloric acid.

b Calculate the percentage of ethanoic acid molecules that have ionised.

3 Explain why sulfuric acid of the same concentration as HCl has a different pH.

ISBN 978 0 6557 0015 9

4 a Write chemical equations for the three ionisation stages of phosphoric acid.

b Considering the triprotic nature of phosphoric acid and the measured pH of a 0.1 M solution, is phosphoric acid a weak or strong acid? Explain your answer.

5 Write equations for the acid and base hydrolysis reaction undergone by each of the amphiprotic anions tested in Part B. In each case, indicate which reaction predominates.

Solution	Formula of anion	Equations	Dominant reaction
0.1 M $NaHCO_3$	HCO_3^-	acid: base:	
0.1 M $NaHSO_4$	HSO_4^-	acid: base:	
0.1 M Na_2HPO_4	HPO_4^{2-}	acid: base:	
0.1 M NaH_2PO_4	$H_2PO_4^-$	acid: base:	

6 Explain your observations in Part C and write an equation for the reaction that occurred.

CONCLUSION

 ISBN 978 0 6557 0015 9

PRACTICAL ACTIVITY 17

Controlled experiment

Reactions of HCl with metals and carbonates

SUGGESTED DURATION

- 50 minutes

INTRODUCTION

Acids show similar chemical behaviour. They react with many metals and metal compounds in common ways meaning predictions can be made about the products.

AIM

To investigate some common reactions of hydrochloric acid

MATERIALS

- 0.1 M HCl
- 1 M HCl
- limewater (saturated calcium hydroxide solution)
- solid samples of $CaCO_3$, $CuCO_3$, $NaHCO_3$, Mg, Al
- test tubes/small beakers
- spatula
- plastic Pasteur pipette
- safety glasses/goggles
- disposable gloves
- taper
- matches

Take note of the safety requirements for hydrogen produced in small quantities.

PRE-LAB SAFETY INFORMATION		
Material used/other risks	**Hazard**	**Control**
0.1 M hydrochloric acid	• non-hazardous < 3 M	Wear safety glasses/goggles.
1 M hydrochloric acid	• non-hazardous < 3 M	Wear safety glasses/goggles.
calcium carbonate	• non-hazardous	
copper(II) carbonate	• harmful if swallowed • causes skin irritation and serious eye irritation • may cause respiratory irritation	Wear safety glasses/goggles, disposable gloves and a lab coat.
magnesium	• on contact with water, releases flammable gases, which may ignite spontaneously	Do not use near open flame, sparks etc.
0.1 M sodium hydroxide	• mild eye irritant	Wear safety glasses/goggles.
sodium hydrogen carbonate	• non-hazardous	
limewater (saturated calcium hydroxide solution)	• non-hazardous	Wear safety glasses/goggles.

Please indicate that you have understood the information in the safety table.

Name (print): ____________________

I understand the safety information (signature): ____________________

METHOD

1 ▪ Put on your PPE, ensuring your lab coat is buttoned, your gloves are fitted correctly and comfortably, and your safety glasses/goggles are fitted over your eyes.

2 ▪ Predict the products of each reaction shown in Table 1 and record the expected products in the appropriate column.

3 ▪ Carry out each of the following tests and record your observations in Table 1. If any gas is evolved, test for hydrogen or carbon dioxide using the two tests shown in steps 4 and 5.

  a Add about 1 mL of 1 M HCl to a test tube containing a small amount of solid $CaCO_3$.

  b Add about 1 mL of 1 M HCl to a small amount of solid CuO in a test tube. Shake the test tube and allow it to stand for 5 minutes.

  c Add about 10 mL of 1.0 M HCl to a test tube containing a 1 cm strip of magnesium.

  d Add about 10 mL of 1.0 M HCl to a test tube containing a 1 cm strip of aluminium.

  e Add 1 mL of 0.1 M HCl to a small amount of solid $NaHCO_3$ in a test tube.

4 ▪ Testing for hydrogen gas: The pop test is used to test for the presence of hydrogen gas. Evolved gas can be collected in an inverted test tube and then tested with a lighted taper. If H_2 gas is present, a loud pop will be heard.

5 ▪ Testing for carbon dioxide: Limewater is used to test for the presence of carbon dioxide. When evolved CO_2 gas is bubbled through limewater, the solution becomes cloudy.

6 ▪ Spray and wipe down your bench, return used equipment, wash equipment as necessary, and remove PPE as directed by your teacher.

RESULTS

Table 1 Predictions and observations for the reactions

Reaction	Predicted products	Observations
hydrochloric acid + calcium carbonate		
hydrochloric acid + copper(II) carbonate		
hydrochloric acid + magnesium		
hydrochloric acid + aluminium		
hydrogen chloride gas + sodium hydrogen carbonate		

DISCUSSION

1 Comment on how closely your observations matched your predictions.

 ISBN 978 0 6557 0015 9

2 Write balanced full and ionic equations for each reaction in Table 2.

Table 2 Balanced full and ionic equations for reactions

Reaction	Equations
hydrochloric acid + calcium carbonate	full: ionic:
hydrochloric acid + copper(II) carbonate	full: ionic:
hydrochloric acid + magnesium	full: ionic:
hydrochloric acid + aluminium	full: ionic:
hydrochloric acid + sodium hydrogen carbonate	full: ionic:

3 Summarise the common reactions of HCl and other compounds you have observed.

CONCLUSION

PRACTICAL ACTIVITY 18

Controlled experiment

Beetroot—a natural indicator

SUGGESTED DURATION

- 50 minutes (Part A = 20 minutes, Part B = 30 minutes)

AIM

To extract a natural indicator from beetroot and test its colours in acidic, neutral and basic solutions. The colours of another natural indicator, litmus, and several synthetic indicators in these solutions are also tested

PRE-LAB SAFETY INFORMATION

Material used/other risks	Hazard	Control
0.1 M hydrochloric acid	• non-hazardous < 3 M	Wear safety glasses/ goggles.
0.1 M sodium hydroxide	• mild eye irritant	Wear safety glasses/ goggles.
beetroot	• stains clothes, hands and benches	Wear disposable gloves and a lab coat.
phenolphthalein	• flammable liquid and vapour • causes eye irritation • may cause cancer • suspected of causing genetic defects • suspected of damaging fertility	Wear safety glasses/ goggle, disposable gloves and a lab coat.
bromothymol blue	• flammable liquid and vapour • causes eye irritation	Wear safety glasses/ goggles.
Please indicate that you have understood the information in the safety table. Name (print): I understand the safety information (signature):		

MATERIALS

- beetroot
- 100 mL deionised water
- food processor
- fine strainer
- 4 test tubes
- 250 mL glass beaker
- 15 mL of the following solutions:
  - 0.1 M sodium chloride, NaCl(aq)
  - 0.1 M hydrochloric acid, HCl(aq)
  - 0.1 M sodium hydroxide, NaOH(aq)
- 3 plastic Pasteur pipettes
- glass stirring rod
- 100 mL glass beaker
- dropper bottles (optional, for Part B) containing:
  - litmus solution
  - phenolphthalein
  - methyl orange
  - bromothymol blue
- pH meter
- pH probe with data logger (optional, for Part B)
- safety glasses/goggles
- disposable gloves

METHOD

Put on your PPE, ensuring your lab coat is buttoned, your gloves are fitted correctly and comfortably, and your safety glasses/goggles are fitted over your eyes.

PART A • COLOUR CHANGES OF INDICATORS

1. Peel and chop the beetroot. Place it in a blender with 100 mL distilled water and blend the mixture until relatively smooth.
2. Using the fine strainer, strain the mixture into a beaker.
3. To each of four test tubes, add about 3 mL of one of the following liquids: water, NaCl solution, HCl solution, NaOH solution.
4. Add three drops of beetroot juice to each of the test tubes and stir. Record your results in Table 1.
5. Rinse and dry the test tubes, refilling with the four solutions, and conduct the test using the next indicator. Repeat for the remaining two indicators.

ISBN 978 0 6557 0015 9

PART B • pH RANGES OF INDICATORS

For beetroot, phenolphthalein and bromothymol blue

1 ▪ Add three drops of indicator to 10 mL of deionised water in a small beaker.

2 ▪ Measure the pH of the solution using a probe and record your measurement in Table 2. Add NaOH solution to the beaker, drop-wise while stirring, until the first colour change is observed. Insert the pH probe and record the pH of this solution.

3 ▪ Add a few more drops of NaOH to determine any further change, until the colour remains constant. Again, insert the pH probe and record the pH of this solution.

For blue litmus and methyl orange

4 ▪ Add three drops of indicator to 10 mL of deionised water in a small beaker.

5 ▪ Measure the pH of the solution using a probe and record your measurement in Table 2. Add HCl drop-wise (instead of NaOH) while stirring, until the first colour change is observed. Insert the pH probe and record the pH of this solution.

6 ▪ Add a few more drops of HCl to determine any further change, until the colour remains constant. Again, insert the pH probe and record the pH of this solution.

7 ▪ Spray and wipe down your bench, return used equipment, wash equipment as necessary, and remove PPE as directed by your teacher.

RESULTS

Table 1 Indicator colours in different solutions

Indicator	Colour in water	Colour in NaCl solution	Colour in HCl solution	Colour in NaOH solution
beetroot				
litmus				
phenolphthalein				
methyl orange				
bromothymol blue				

Table 2 Indicator pH ranges

Indicator	Initial pH	pH at start of colour change	pH at definite colour change
beetroot			
litmus			
phenolphthalein			
methyl orange			
bromothymol blue			

DISCUSSION

1 State whether beetroot juice is suitable as an indicator to identify the following.

a acidic substances

b basic solutions

Give reasons for your answers.

2 Use your results to state the pH range over which the beetroot indicator changes colour.

3 From your experimental observations, or from secondary sources, which of the following indicators—beetroot juice, phenolphthalein, blue litmus, methyl orange and bromothymol blue—would be effective in distinguishing between the following?

a NaCl and HCl solutions

b NaCl and NaOH solutions

4 From your observations, or from secondary sources, find out the pH range and colour change for the following indicator.

a phenolphthalein

b methyl orange

c bromothymol blue

CONCLUSION

 ISBN 978 0 6557 0015 9

PRACTICAL ACTIVITY 19

Controlled experiment

Reactivity series of metals

SUGGESTED DURATION

- 50 minutes

INTRODUCTION

In general, ions of metal X in solution will be converted to the metal (that is, displaced from the solution) by another metal that is more easily oxidised than metal X. The more easily a metal is oxidised, the more reactive it is.

The results of displacement reactions between metals and metallic ions in solution can be used to arrange metals in order of their ability to be oxidised (or the ability of their ions to be reduced) to form a reactivity series.

AIM

To obtain a reactivity series of metals by investigating reactions in which metals displace other metal ions from solutions

MATERIALS

- 40 mL of 0.1 M magnesium nitrate solution, $Mg(NO_3)_2$
- 40 mL of 0.1 M copper(II) sulfate solution, $CuSO_4$
- 40 mL of 0.1 M zinc sulfate solution, $ZnSO_4$
- 40 mL of 0.1 M iron(II) sulfate, $FeSO_4$
- 4 pieces of magnesium, Mg, approx. 2 cm long
- 4 pieces of copper, Cu, approx. 1 cm^2
- 4 pieces of zinc, Zn, approx. 1 cm^2
- 4 iron nails or iron wire, Fe
- emery paper (sandpaper with fabric backing)
- 16 test tubes
- test-tube rack
- safety glasses/goggles
- disposable gloves

PRE-LAB SAFETY INFORMATION

Material used/other risks	Hazard	Control
0.1 M magnesium nitrate	• non-hazardous	
0.1 M copper(II) sulfate	• causes mild skin irritation • toxic to aquatic life with long-lasting effects	Wear safety glasses/goggles, disposable gloves and a lab coat.
0.1 M iron(II) sulfate	• causes mild skin irritation	Wear safety glasses/goggles, disposable gloves and a lab coat.
0.1 M zinc sulfate	• causes serious eye irritation • very toxic to aquatic life with long-lasting effects	Wear safety glasses/goggles.
magnesium ribbon	• in contact with water, releases flammable gases, which may ignite spontaneously	Do not use near open flame, sparks/etc.
zinc sheet	• very toxic to aquatic life with long-lasting effects	

Please indicate that you have understood the information in the safety table.

Name (print): ______________________

I understand the safety information (signature): ______________________

METHOD

1 ▪ Put on your PPE, ensuring your lab coat is buttoned, your gloves are fitted correctly and comfortably, and your safety glasses/goggles are fitted over your eyes.

2 ▪ Clean a sample of each of the four metals using emery paper. Place each sample in a test tube.

3 ▪ Add approximately 5 mL of magnesium nitrate solution to each of the test tubes.

4 ▪ Leave them to stand for about 5 minutes. Record any changes that occur.

5 ▪ Repeat steps 2–4 using each of the other solutions in turn. Complete Table 1. Draw a cross (✗) where there was no observable reaction. Draw a tick (✓) and record your observations when there was an observable reaction.

6 ▪ Spray and wipe down your bench, return used equipment, wash equipment as necessary, and remove PPE as directed by your teacher.

RESULTS

Table 1 Observations of reactions between metals and metal ions

Metal	Metal ion			
	Mg^{2+}	Cu^{2+}	Zn^{2+}	Fe^{2+}
Mg				
Cu				
Zn				
Fe				

DISCUSSION

1 a Which metal was the most reactive?

b Which metal was the least reactive?

c List the metals in order from most reactive to least reactive.

2 Write half-equations and full equations for the redox reactions that took place during this experiment.

 ISBN 978 0 6557 0015 9

3 List the reduction half-equations as a reactivity series.

4 Compare the order of your equations with that of the equations listed in the Key Knowledge section (Figure 2.1.9 on page 115). Why might differences arise?

CONCLUSION

PRACTICAL ACTIVITY 20

Classification and identification

Comparing a simple primary cell and a direct reaction

SUGGESTED DURATION

- 30 minutes

INTRODUCTION

A primary cell is a galvanic cell in which chemical energy is converted to electrical energy. The Daniell cell was invented in 1836 by John Frederic Daniell, a British chemist and meteorologist. A Daniell cell consists of Cu/Cu^{2+} and Zn/Zn^{2+} half-cells.

AIM

To construct a simple primary cell and experimentally determine the voltage produced

As directed by your teacher, complete the pre-lab safety information by referring to the safety data sheets (SDSs) or your teacher's risk assessment for the activity. 

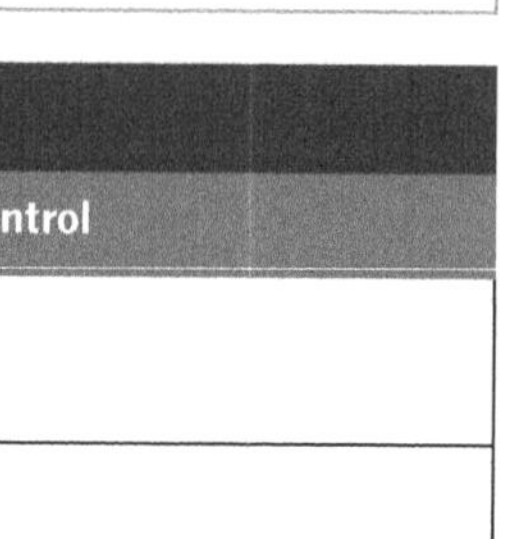

PRE-LAB SAFETY INFORMATION		
Material used/other risks	**Hazard**	**Control**
0.1 M copper(II) nitrate solution		
0.1 M zinc nitrate solution		
0.1 M potassium nitrate solution		
zinc strips		

Complete and indicate that you have understood the information in the safety table.

Name (print):

I understand the safety information (signature):

MATERIALS

- 100 mL of 0.1 M copper(II) nitrate solution, $Cu(NO_3)_2$
- 50 mL of 0.1 M zinc nitrate solution, $Zn(NO_3)_2$
- 50 mL of 0.1 M potassium nitrate solution, KNO_3
- copper electrode, Cu
- 2 zinc strips, Zn
- 4 × 100 mL glass beakers
- millivolt meter (mV) or multimeter
- 2 wire leads with alligator clips
- 1 strip of 2 cm × 15 cm chromatography paper, cartridge paper or filter paper
- safety glasses/goggles
- disposable gloves

METHOD

1. Put on your PPE per the risk assessment and management table you filled in. As relevant, ensure your lab coat is buttoned, your gloves are fitted correctly and comfortably, and your safety glasses/goggles are fitted over your eyes.
2. Add 50 mL of copper(II) nitrate to a 100 mL beaker and add a strip of zinc. Leave for 10 minutes then record your observations.
3. Construct the following half-cells.
   a. Cu^{2+}/Cu: a 100 mL beaker containing a copper electrode and 50 mL of copper(II) nitrate solution
   b. Zn^{2+}/Zn: a 100 mL beaker containing a zinc strip and 50 mL of zinc nitrate solution.
4. Make a salt bridge by soaking a strip of filter paper in a beaker of potassium nitrate solution.
5. Join the Cu^{2+}/Cu and Zn^{2+}/Zn half-cells using the salt bridge. Attach two wires to the voltmeter and clip the other end of one wire to the copper electrode. Momentarily touch the loose end of the second wire to the zinc electrode. If the needle of the voltmeter is deflected onto the scale, clip this loose end of the wire to the zinc electrode and proceed with the next step. If the needle of the voltmeter is deflected below zero, swap the wires at the terminals of the voltmeter before continuing.

ISBN 978 0 6557 0015 9

6 ▪ Record the voltage and identify the positive and negative electrodes. (The positive electrode is connected to the positive terminal of the voltmeter.)

7 ▪ Spray and wipe down your bench, return used equipment, wash equipment as necessary, and remove PPE as directed by your teacher.

RESULTS

Observations of zinc strip in copper(II) nitrate solution:

Table 1 Voltage and electrode polarity

Cell	Voltage (mV)	Positive electrode
Cu^{2+}/Cu and Zn^{2+}/Zn		

DISCUSSION

1 What can you infer about the reaction occurring in the first beaker containing copper(II) nitrate and a zinc strip?

2 Draw a diagram of the Daniell cell you constructed. Label the contents of the half-cells, the positive and negative electrodes, the anode and cathode, the direction of electron flow through the external circuit and the direction the ions move in the salt bridge.

3 Write half-equations for the reactions occurring in each half-cell and hence write an overall equation for the reaction in each cell.

Cu/Cu^{2+} half-cell:

Zn/Zn^{2+} half-cell:

Overall equation:

ISBN 978 0 6557 0015 9

4 What is the function of the salt bridge in the Daniell cell?

5 With reference to your observations of the first beaker, describe the aspects of the Daniell cell that make it able to produce electricity.

CONCLUSION

 ISBN 978 0 6557 0015 9

EXAM-STYLE QUESTIONS

Multiple-choice questions

Question 1

What percentage of water on Earth is available for drinking?

A. less than 1%

B. 2.5%

C. 5%

D. 7.5%

Question 2

Which one of the following correctly identifies the interaction between different molecules of the listed group 16 hydrides?

	H_2O	H_2S
A	covalent bonding	dipole–dipole interaction
B	hydrogen bonding	dipole–dipole interaction
C	dipole–dipole interaction	covalent bonding
D	hydrogen bonding	hydrogen bonding

Question 3

Which one of the following best defines an acid according to the Brønsted–Lowry theory?

A. a substance that donates H^+ ions

B. a substance that reacts to produce OH^- ions

C. a substance that produces H^+ ions in water

D. a substance that contains protons

Question 4

Which equation represents an acid–base reaction?

A. $Zn(s) + 2HCl(aq) \longrightarrow ZnCl_2(aq) + H_2(g)$

B. $HNO_3(aq) + NH_3(aq) \longrightarrow NH_4^+(aq) + NO_3^-(aq)$

C. $KCl(aq) + AgNO_3(aq) \longrightarrow AgCl(s) + KNO_3(aq)$

D. $FeCl_2(aq) + Cl_2(g) \longrightarrow 2FeCl_3(aq)$

Question 5

If solution A has a pH of 1 and solution B has a pH of 2, the ratio of the concentration of hydrogen ion in solution A to that in solution B is:

A. 1 : 2

B. 2 : 1

C. 1 : 10

D. 10 : 1

Question 6

The concentrations of ions in four different solutions are listed below at 25°C. Which one is acidic?

A. solution I, $[H^+] = 10^{-7}$ M

B. solution II, $[OH^-] = 10^{-5}$ M

C. solution III, $[H^+] = 10^{-9}$ M

D. solution IV, $[OH^-] = 10^{-8}$ M

EXAM-STYLE QUESTIONS

Question 7

Identify the products of the reaction between nitric acid, HNO_3, and magnesium carbonate.

A. magnesium nitrate and water

B. magnesium nitrate and carbon dioxide

C. magnesium nitrate, carbon dioxide and water

D. magnesium nitrate, carbon dioxide and hydrogen gas

Question 8

Which of the following will undergo a spontaneous reaction when added to sodium nitrate solution?

A. zinc nitrate solution

B. calcium sulfate powder

C. potassium metal

D. zinc metal

Question 9

When silver nitrate reacts with sodium chloride, a precipitate forms according to the equation:

$AgNO_3(aq) + NaCl(aq) \longrightarrow AgCl(s) + NaNO_3(aq)$

Identify the spectator ions.

A. Ag^+ and Cl^-

B. Na^+ and Cl^-

C. Na^+ and NO_3^-

D. Ag^+ and NO_3^-

Question 10

Which list has the relevant metals in order of increasing reactivity?

A. Au < Zn < Ca < K

B. Au < Ca < Mg < Na

C. Pb < Au < Ca < Zn

D. K < Na < Ca < Au

Short-answer questions

Question 1

Some people pay a lot of money to have their bodies frozen when they die. They hope that they will be brought back to life sometime in the future, perhaps when cures for diseases are known or when recipes for longevity are discovered. Evaluate why the blood of these people is replaced with another liquid before their bodies are frozen. 3 marks

ISBN 978 0 6557 0015 9

EXAM-STYLE QUESTIONS

Question 2

Write a full balanced chemical equation, including states, to represent each of the following reactions. 6 marks

a. HCl gas is bubbled through water.

b. Solid magnesium reacts with hydrochloric acid, HCl(aq).

c. Nitric acid, HNO_3, is added to potassium carbonate solution.

Question 3

Beaker A contains 100 mL of 0.10 M ethanoic acid. Beaker B contains 100 mL of 0.10 M hydrochloric acid.

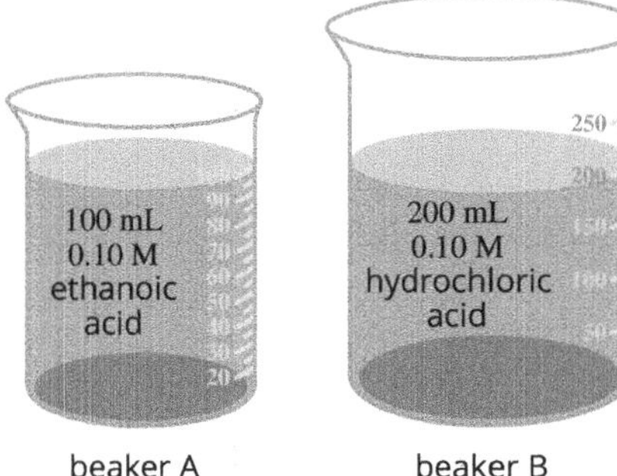

a. i. Calculate the pH of the solution in beaker B. 1 mark

ii. Will the solution in beaker A have a higher of lower pH than the solution in beaker B? Justify your answer. 4 marks

b. A student uses beetroot indicator and universal indicator to measure the pH of beaker B. Which of these will give the most accurate result? Explain your answer. 2 marks

c. Write a full equation for the reaction between ethanoic acid, CH_3COOH, and sodium hydroxide, NaOH. 1 mark

EXAM-STYLE QUESTIONS

Question 4

A reduction in the quality of the atmosphere can have significant consequences for Earth's inhabitants. One such problem is ocean acidification.

a. Describe how an increase in CO_2 in the atmosphere can lead to ocean acidification. Include an equation in your answer. 3 marks

b. Describe one example of ocean acidification having an effect on marine organisms. Include an equation in your answer. 2 marks

Question 5

A student investigates the reactivity of metals A, B, C and D by combining them with metal ions in solution and observing whether a reaction occurs. The results of the student's investigation are shown in the table. A tick indicates a reaction was observed. A cross indicates there was no reaction.

TABLE 1 Results of reactions between metals and metal ions

Metal ion \ Metal	A	B	C	D
A^+		✗	✓	✗
B^{2+}	✓		✓	✗
C^{3+}	✗	✗		✗
D^+	✓	✓	✓	

a. Place the metals in order of reactivity from most reactive to least reactive. 1 mark

b. Write a half-equation for the reaction of the strongest oxidising agent. 1 mark

c. Identify the equation you wrote in part **b** as oxidation or reduction. 1 mark

 ISBN 978 0 6557 0015 9

UNIT 2

How do chemical reactions shape the natural world?

AREA OF STUDY 2

How are chemicals measured and analysed?

Outcome

On completion of this unit the student should be able to calculate solution concentrations and predict solubilities, use volumetric analysis and instrumental techniques to analyse for acids, bases and salts, and apply stoichiometry to calculate chemical quantities.

Key knowledge

Measuring solubility and concentration

- solution concentration as a measure of the quantity of solute dissolved in a given mass or volume of solution (mol L^{-1}, g L^{-1}, %(m/v), %(v/v), ppm), including unit conversions
- the use of solubility tables and solubility graphs to predict experimental determination of ionic compound solubility; the effect of temperature on the solubility of a given solid, liquid or gas in water
- the use of precipitation reactions to remove impurities from water

Analysis for acids and bases

- volume–volume stoichiometry (solutions only) and application of volumetric analysis, including the use of indicators, calculations related to the preparation of standard solutions, dilution of solutions, and use of acid–base titrations (excluding back titrations) to determine the concentration of an acid or a base in a water sample

Measuring gases

- CO_2, CH_4 and H_2O as three of the major gases that contribute to the natural and enhanced greenhouse effects due to their ability to absorb infrared radiation
- the definitions of gas pressure and standard laboratory conditions (SLC) at 25°C and 100 kPa
- calculations using the ideal gas equation ($pV = nRT$), limited to the units kPa, Pa, atm, mL, L ,°C and K (including unit conversions)
- the use of stoichiometry to solve calculations related to chemical reactions involving gases (including moles, mass and volume of gases)
- calculations of the molar volume or molar mass of a gas produced by a chemical reaction

Analysis for salts

- sources of salts found in water or soil (which may include minerals, heavy metals, organo-metallic substances) and the use of electrical conductivity to assess the salinity and quality of water or soil samples
- quantitative analysis of salts:
  - molar ratio of water of hydration for an ionic compound
  - the application of mass–mass stoichiometry to determine the mass present of an ionic compound
  - the application of colorimetry and/or UV–visible spectroscopy, including the use of a calibration curve to determine the concentration of ions or complexes in a water or soil sample

Measuring solubility and concentration

Different solutions can contain different amounts of dissolved solutes. In many contexts, it is important for chemists to know how much solute can dissolve in a particular volume of solution, and what the concentration of the solution is once it has dissolved.

CONCENTRATION OF SOLUTIONS

Solutes dissolve in solvents to form solutions. The **concentration** of a solution is a measure of the number of moles or mass of solute dissolved in a given mass or volume of solution. A number of different concentration units are used in various domestic, environmental, commercial or industrial contexts (Table 2.2.1).

Table 2.2.1 Different measures used for expressing the concentration of a solution

Units	Description	Example of use
M or mol L^{-1}	amount, in mol, of solute per litre of solution	Chemicals in a laboratory: 1.0 M HCl, 0.10 M NaOH.
g L^{-1}	mass of solute, in g, per litre of solution	A bottle of bleach gives the concentration of sodium hypochlorite as 56 g L^{-1}.
mg L^{-1}	mass of solute, in mg, per litre of solution	Melbourne's water supply has calcium ion concentrations typically 5–10 mg L^{-1}.
% (m/v)	percentage mass by volume—describes the mass of solute, in g, per 100 mL of solution	Saline drips are a solution of 0.9% (m/v) sodium chloride.
% (v/v)	percentage by volume—describes the volume of solute, in mL, per 100 mL of solution	Alcohol content of wine might be 13.5% (v/v).
ppm	parts per million, e.g. mass, in g, per million grams, equivalent to mg kg^{-1}, or mg L^{-1}	The legal limit of mercury, for safe human consumption, is 0.5 ppm for most species of Australian fish.

This means the concentration of any solution can be expressed in a number of different ways (Figure 2.2.1).

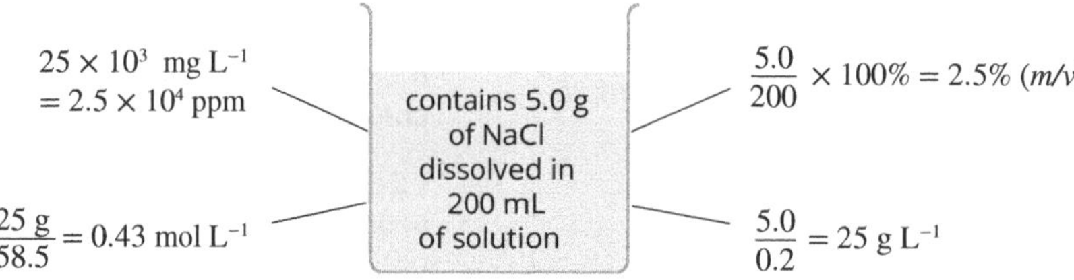

Figure 2.2.1 Different ways of expressing the concentration of a solution

Some useful relationships for converting between different units of solution are:

- $1 \text{ g} = 10^3 \text{ mg} = 10^6 \mu\text{g}$
- 1% (m/v) $= 10 \text{ g L}^{-1}$
- $1 \text{ ppm} = 1 \text{ g}/10^6 \text{ g} = 1 \text{ mg kg}^{-1} = 1 \text{ mg L}^{-1}$

Molar concentration

Chemists often measure concentration in terms of molarity, which is the amount, in mol, of a solute dissolved per litre of solution (mol L^{-1}). Molarity is given the symbol c and the unit M. A solution with a molarity of 5 mol L^{-1} is said to be a '5 molar' solution and its concentration can also be written as '5 M'. A useful formula links the amount of solute dissolved (n), the concentration of a solution (c) and the volume of the solution (V).

volume, in L

amount in mol —— $n = cV$

concentration, in M

DILUTION

Dilution of a solution by adding water changes the concentration of the solution (Figure 2.2.2).

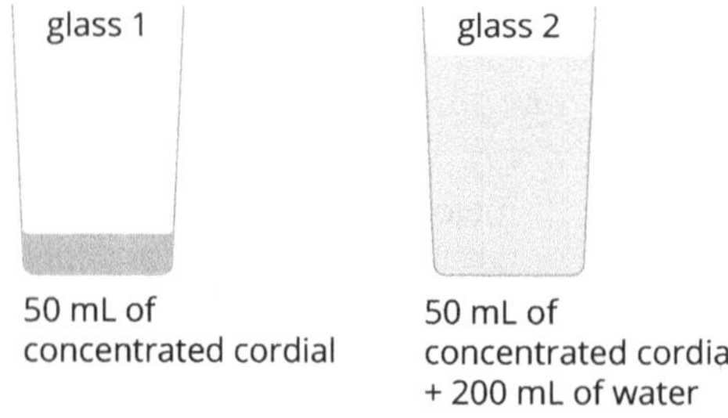

Figure 2.2.2 When cordial is diluted with water, the amount of solute does not change. However, the solute particles are more widely spaced in a greater volume of solvent, so the concentration has decreased.

The amount of solute does not change, so $n_1 = n_2$. A useful relationship for calculating the concentration of the diluted solution is:

$$c_1V_1 = c_2V_2$$

 ISBN 978 0 6557 0015 9

In this equation, c_1 and V_1 are the concentration and volume of the initial solution, and c_2 and V_2 are the concentration and volume of the diluted solution.

SOLUBILITY

The **solubility** of a substance is a measure of the mass of solute in grams that can dissolve per 100 g of water. Where concentration is a measure of how much solute is actually dissolved per unit of solution, solubility is a measure of the maximum amount of solute that could potentially dissolve per 100 g of water.

SATURATION OF A SOLUTION

A solution in which the maximum amount of solute is dissoved at that particular temperature is called a **saturated solution**. An **unsaturated solution** is one in which more solute can be dissolved than is currently present. A **supersaturated solution** is one in which more solute is dissolved than can usually be dissolved at that temperature under normal conditions.

Not all ionic substances dissolve well in water, as you learnt in Unit 1 Area of Study 1, although the ability of solids to dissolve tends to increase with increasing water temperature.

A **solubility table** summarises the solubility in water of many common ionic compounds. However, it is useful to remember that:

- all compounds containing a nitrate ion are soluble
- all compounds of group 1 elements are soluble.

Solubility tables were covered in Unit 1 Area of Study 1. You can find a solubility table in Table 1.1.18 on page 13.

SOLUBILITY CURVES

The solubility of many substances in water varies with changing temperature. In general, the:

- solubility of solids and liquids increases with increasing temperature
- solubility of gases decreases with increasing temperature.

The quantitative relationship between solubility and temperature can be represented by a **solubility curve** (Figure 2.2.3). The curve can be used to predict ionic compound solubility at any temperature.

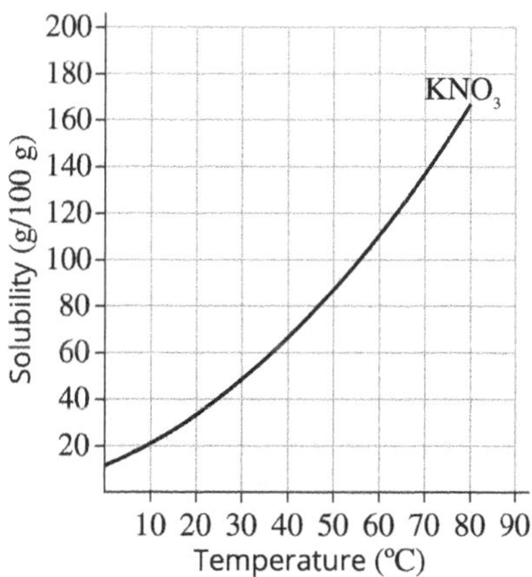

Figure 2.2.3 The solubility curve for potassium nitrate

Each point on the solubility graph represents a **saturated solution**—no more solute can be dissolved at that particular temperature. Any point below the line represents an **unsaturated solution**. An unsaturated solution contains less solute than is needed to make a solution saturated at that temperature.

PRECIPITATION REACTIONS CAN REMOVE IMPURITIES FROM WATER

An understanding of the solubility of different ionic substances allows precipitation reactions to be used to remove impurities from water, such as when water is purified for drinking. Precipitation reactions were covered in Unit 1 Area of Study 1 on page 13.

For example, heavy metal ions such as Pb^{2+}, Cu^{2+} and Ni^{2+} may be present in wastewater from industries. An effective way to remove them from solution is to add suitable anions to form insoluble compounds. The precipitates formed can then be filtered out of the solution. $Ca(OH)_2$ added to water will precipitate Pb^{2+} ions according to the equation:

$$Pb^{2+}(aq) + 2OH^-(aq) \rightarrow Pb(OH)_2(s)$$

- **You will now be able to complete Worksheets 33–36 and conduct Practical activity 21.**

Analysis for acids and bases

The ratio provided by the coefficients in a balanced chemical equation is also a **mole ratio** by which atoms or molecules react and are produced. The study of the ratios of moles is called **stoichiometry**.

Consider the reaction represented by the following chemical equation:

$$Mg(s) + 2HCl(aq) \rightarrow MgCl_2(aq) + H_2(g)$$

From this equation it can be seen that:

- 2 mol HCl is required for the complete reaction of 1 mol Mg
- 1 mol Mg and 2 mol HCl are needed to react to produce 1 mol $MgCl_2$
- the mole ratio Mg : HCl is 1:2.

VOLUME-VOLUME STOICHIOMETRY

Volume–volume stoichiometry allows calculation of the concentration of an unknown solution. It uses the volumes of both reactants as well as the concentration of one reactant to determine the concentration of the other reactant. The coefficients in the balanced chemical equation for the reaction provide the mole ratio used for stoichiometry.

Consider the precipitation reaction represented by the following chemical equation:

$$2HCl(aq) + Na_2CO_3(aq) \rightarrow 2NaCl(aq) + H_2O(l) + CO_2(g)$$

From this equation it can be seen:

- 2 mol of HCl is required for the complete reaction of 1 mol of Na_2CO_3
- 2 mol of HCl and 1 mol of Na_2CO_3 will produce 2 mol NaCl, 1 mol H_2O and 1 mol CO_2
- the mole ratio NaCl : Na_2CO_3 is 2:1.

KEY KNOWLEDGE

Determining the concentration of an unknown solution from the concentration and volume of another solution requires calculation of the number of moles, so the mole ratio can be used (Figure 2.2.4).

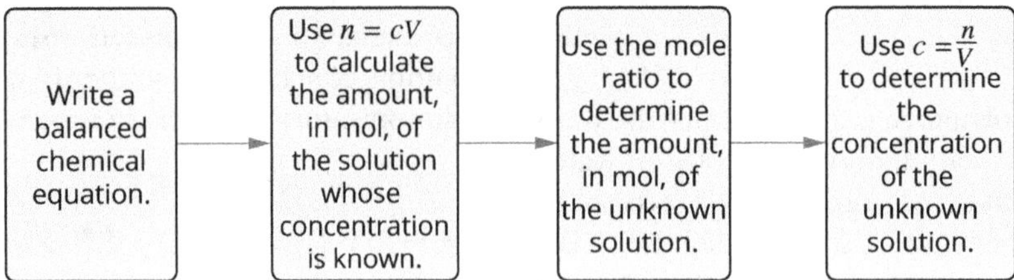

Figure 2.2.4 A flow chart for volume–volume stoichiometric calculations

***Example*: Determine the concentration of HCl in mol L^{-1} that is present in a 25.00 mL sample that completely reacts with 19.45 mL of 0.1005 M Na_2CO_3.**

Thinking	Working
1 Calculate the amount, in mol, of the known solution using $n = cV$.	$n(Na_2CO_3) = 0.1005 \times 0.01945$ $= 0.001\,955$ mol
2 Use the mole ratio from the equation to calculate the amount, in mol, of product.	$\frac{n(HCl)}{n(Na_2CO_3)} = \frac{2}{1}$ So $n(HCl) = \frac{2}{1} \times n(Na_2CO_3)$ $= \frac{2}{1} \times 0.001\,955$ $= 0.003\,909$ mol
3 Determine the concentration of the unknown solution.	$c(HCl) = \frac{n}{V}$ $= \frac{0.003\,909}{0.025\,00}$ $= 0.1564$ M

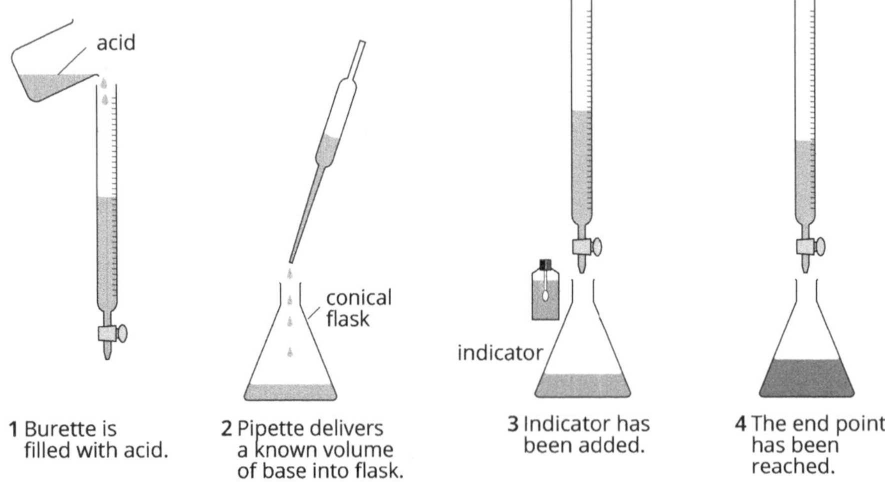

Figure 2.2.5 A simple titration

VOLUMETRIC ANALYSIS

Acid–base titrations can be used to determine the concentration of an acid or a base in a water sample. Acid–base titrations are a type of volumetric analysis and require the use of volume–volume stoichiometry.

Acid-base titration

In a simple acid–base titration (Figure 2.2.5), one solution is dispensed from a burette into a conical flask that contains a volume of another solution, which has been precisely measured using a pipette (Table 2.2.2). An acid–base **indicator** is added to the conical flask, selected so that it changes colour when stoichiometric amounts of reactants have been added, making it possible to see when the reaction is complete.

 ISBN 978 0 6557 0015 9

KEY KNOWLEDGE

Table 2.2.2 Summary of the terms used in volumetric analysis

Term	Description
primary standard	A substance so pure that the amount can be accurately calculated from its mass, e.g. Na_2CO_3. For a chemical to be a primary standard, it should be easily obtained, have a known formula and be easily stored without reacting with water or gases in the atmosphere.
standard solution	A solution with an accurately known concentration. This illustration shows a standard solution being made from a primary standard. 1 Place weighed sample in volumetric flask. 2 Half-fill with water. Shake to dissolve the sample. 3 Add water to the calibration line. Shake again.
burette	a calibrated glass tube that delivers variable volumes accurately
titre	the volume delivered by a burette
pipette	a calibrated glass tube that delivers a fixed volume accurately
aliquot	the volume delivered by a pipette
end point	the point at which the indicator changes colour
equivalence point	the point in the titration where the stoichiometric proportions of the reactants have been mixed
indicator	a weak acid whose conjugate base is a different colour

Use of indicators

Acid–base indicators, which you learnt about in Area of Study 1, are weak acids whose conjugate base is a different colour. The colour of an indicator in solution depends on the relative concentrations of the acid and base, and therefore depends on the pH of the solution. The end point of an indicator is the pH at which it changes colour. Table 2.2.3 lists the sources of some common indicators and their pH ranges (the pH over which they change colour).

A suitable indicator is carefully chosen for a specific acid–base titration so that its end point best matches the otherwise colourless equivalence point of the acid–base reaction.

CALCULATIONS IN VOLUMETRIC ANALYSIS

Volumetric analysis involves calculations related to the preparation of standard solutions, dilution of solutions and use of acid–base titrations.

Calculations: preparation of standard solutions

The concentration, in $mol\,L^{-1}$, of a standard solution prepared from a primary standard can be calculated by following the steps outlined in Figure 2.2.6 on page 156.

Table 2.2.3 Some indicators and their origin, colour and pH range

Indicator	Origin	Colour in acid	Colour in base	pH range
phenolphthalein	synthetic	colourless	pink	8.2–10.0
methyl orange	synthetic	pink	yellow	3.2–4.4
bromothymol blue	synthetic	yellow	blue	6.0–7.6

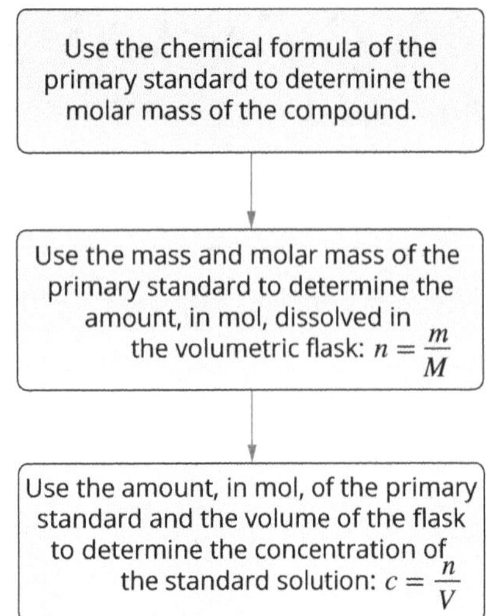

Figure 2.2.6 A flow chart of the steps used in the calculation of the concentration of a standard solution

Calculations: titrations

The calculations used in acid–base titrations use volume–volume stoichiometry, as was outlined in Fig 2.2.4 on page 154. The average titre will be the volume of one solution and the aliquot volume will be the volume of the other.

Dilutions

It is often necessary to dilute a solution before carrying out a titration in order to obtain more convenient titre volumes. In these titrations, a dilution factor must be taken into account.

A 25.00 mL sample diluted to 250.0 mL will have a dilution factor of $\frac{250}{25} = 10$. The concentration determined by titration using the diluted sample will need to be multiplied by 10 to gain the undiluted concentration.

- **You will now be able to complete Worksheets 37 and 38 and conduct Practical activities 22 and 23.**

Measuring gases

Greenhouse gases

A greenhouse gas is a gas that can absorb infrared radiation. Carbon dioxide (CO_2), methane (CH_4) and water vapour (H_2O) are major greenhouse gases.

The **greenhouse effect** is a natural process that keeps Earth's surface warm enough to support life. Energy from the Sun is absorbed by Earth, causing it to warm up, and then energy radiates back towards space in the form of infrared radiation. Greenhouse gases in the atmosphere trap some of that infrared radiation, keeping the Earth's surface temperature relatively warm and stable, and suitable for human habitation.

The enhanced greenhouse effect

Human activities such as burning fossil fuels for energy increase the amount of H_2O and CO_2 in the atmosphere. CH_4 is emitted through agricultural practices and by decaying organic waste in landfills. Higher levels of these major greenhouse gases in the atmosphere contribute to an **enhanced greenhouse effect** that causes warming beyond that which naturally occurs. As the amount of greenhouse gases increases, more heat is trapped (Figure 2.2.7).

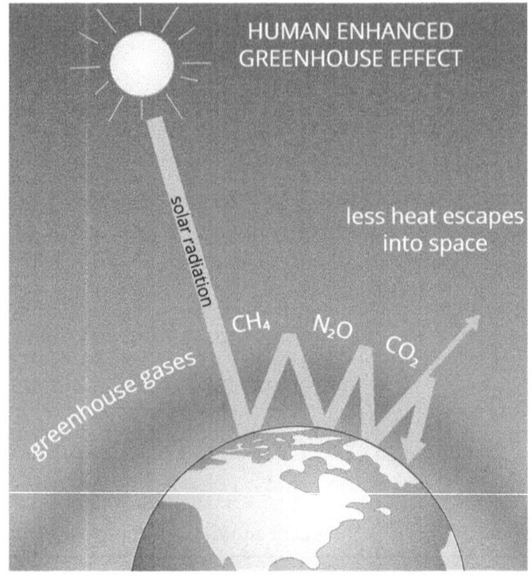

Figure 2.2.7 Human activities are increasing the amount of greenhouse gases in the atmosphere, leading to an enhanced greenhouse effect.

Properties of gases

Gases have the following physical properties. They:

- have low density
- fill all available space
- are easily compressed
- mix rapidly.

The chemical properties of gases depend on the nature of the molecules or atoms on the gas. However, physical properties do not change for different gases.

Units for gas measurements

Physical properties of gases that can be measured include:

- amount
- pressure
- volume
- temperature.

The **amount** of a gas is measured in terms of moles, where 1 mol of a gas contains Avogadro's constant of gas particles. The symbol n is used for amount, with the unit mol.

Pressure is given the symbol p. Pressure is defined as the force exerted on a unit area of substance.

 ISBN 978 0 6557 0015 9

KEY KNOWLEDGE

A number of different units are used to measure pressure (Table 2.2.4).

Table 2.2.4 Equivalent units of pressure

atm	kPa	Pa
1.00	101.3	101 300

Gas volume, symbol V, is the space occupied by a gas. Volume is commonly measured in mL or L. 1 L = 10^3 mL.

When used in calculations, temperatures must be stated in kelvin, rather than in degrees Celsius. A temperature in degrees Celsius (t) can be converted to the temperature in kelvin (T) by using the formula $T = t + 273$ (Table 2.2.5).

Table 2.2.5 Temperatures expressed in degrees Celsius and kelvin

t (°C)	T (K)
−273	0
0	273
100	373

Standard laboratory conditions (SLC) are a standard set of pressure and temperature conditions. At SLC, temperature is 25°C and pressure is 100 kPa.

Ideal gas equation

The ideal gas equation provides a relationship between the pressure, volume, amount and temperature of a gas:

$$pV = nRT$$

The value of R, the ideal gas constant, depends on the units used for pressure, volume and temperature. $R = 8.31\ \text{J K}^{-1}\ \text{mol}^{-1}$ when:

- p is the pressure in kPa
- V is the volume in L
- temperature, T, is stated in kelvin.

The amount, in mol, of a gas at any conditions of volume, pressure and temperature can be calculated by rearranging the equation as:

$$n = \frac{pV}{RT}$$

Molar volume

The molar volume (V_m) of a gas is the volume occupied by one mole of the gas. This is almost the same for different gases at the same temperature and pressure. At SLC, the molar volume of a gas is 24.8 L.

A useful relationship links the amount, in mol, of a gas (n), the volume of the gas (V) and molar volume (V_m):

$$n = \frac{V}{V_m} \quad \text{or} \quad V = n \times V_m$$

STOICHIOMETRY INVOLVING GASES

The mole ratio in a balanced chemical equation can be used for mass–volume or volume–volume stoichiometry involving gases.

Stoichiometric calculations that follow the same general pattern can also be used to calculate the volumes of oxygen required to burn fuels and the volumes of gases produced in any reaction. This is summarised in Figure 2.2.8.

If the reactants and products are in the gaseous state, stoichiometric calculations do not always require the calculation of the number of moles. The same number of moles of different gases occupies the same volume at constant temperature and pressure, so the mole ratio in a balanced chemical equation can also be used as a volume ratio.

CALCULATIONS OF MOLAR VOLUME OR MOLAR MASS OF A GAS

Experimental data obtained in a laboratory can be used to determine the molar volume or molar mass of a gas produced by a chemical reaction.

Molar volume can be determined by using a chemical reaction to produce and measure a volume of gas. The amount, in mol, of gas produced is determined from the amount, in mol, of the limiting reactant, using stoichiometry. The volume of one mole of the gas is determined from:

$$V \text{ of one mole of gas} = \frac{\text{volume of gas produced (L)}}{n \text{ of gas produced (mol)}}$$

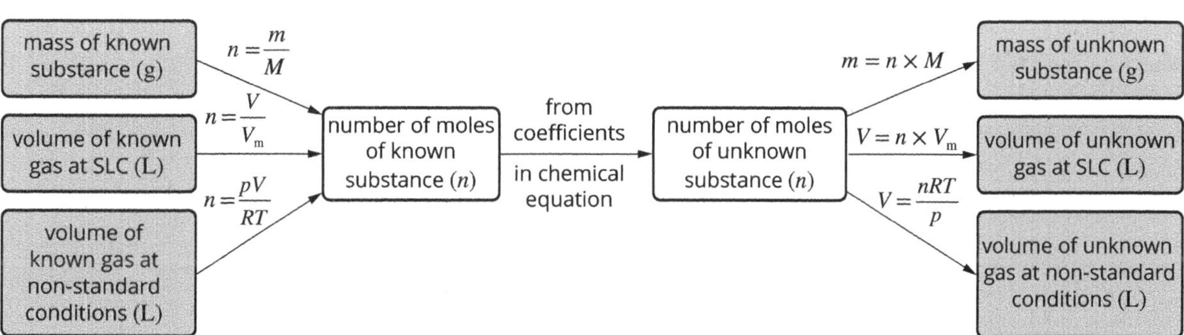

Figure 2.2.8 Stoichiometric calculations generally follow the steps shown in the flow chart. Calculating the number of moles and using a mole ratio from a balanced chemical equation are always central to any stoichiometric calculation.

Experimental conditions of temperature and pressure are obtained. These can be used to determine the molar volume at SLC (25°C and 1 kPa) using:

$$\frac{p_1V_1}{T_1} = \frac{p_2V_2}{T_2}$$

where p_1, V_1 and T_1 are from experimental data, $p_2 = 1\,\text{kPa}$ and $T_2 = 298\,\text{K}$.

Molar mass can be determined by measuring the mass and volume of a gas. The ideal gas equation can be used to find the amount, in mol, of gas in the measured volume at the measured temperature and pressure by using:

$$n = \frac{pV}{RT}$$

Molar mass, in g mol^{-1}, can then be calculated by using:

$$M = \frac{m}{n}$$

- **You will now be able to complete Worksheets 39 and 40 and conduct Practical activities 24 and 25.**

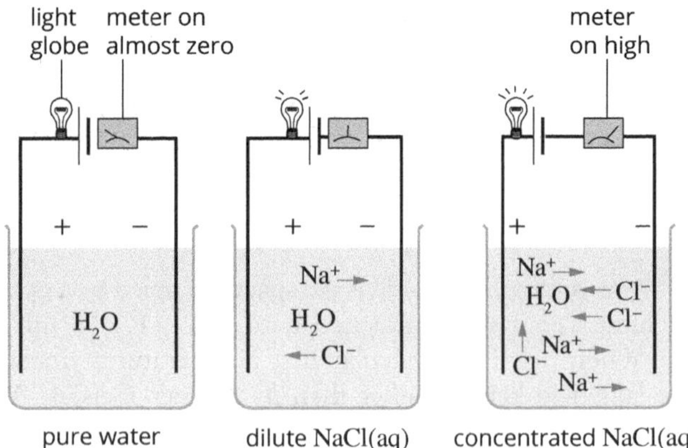

Figure 2.2.9 The conductivity of a salt solution depends on the concentration of salt in the solution. The flow of current increases as the concentration of ions increases.

Analysis for salts

As you have already seen, many salts are very soluble in water. Different sources of water and different soils contain different types and concentrations of salts, which come from a range of sources (Table 2.2.6).

TESTING FOR SALINITY

Salinity is a measure of the total dissolved salts in water or soil. The level of salinity of water samples can be estimated by measuring the electrical conductivity of a water sample. Pure water is a poor conductor of electricity because it contains few ions. As salinity increases, the ion concentration increases and the ability to conduct electricity increases (Figure 2.2.9). The salinity of a water or soil sample may be contributed to by a range of different salts, so this method of analysis gives a measure of total dissolved salts and is an indicator of water quality.

QUANTITATIVE ANALYSIS OF SALTS

Other analysis techniques are required to determine the type and amount of specific salts that are present in water.

Determining molar ratio of water of hydration

A number of ionic salts can become hydrated. This occurs when the ions in the solid salt attract water molecules, which can become enclosed within the crystal lattice. The formula of a hydrated salt can be determined by measuring the mass of water present through evaporation. The mole ratio of water to salt can then be determined.

MASS–MASS STOICHIOMETRY

Mass–mass stoichiometry allows calculation of the mass of reactants consumed and products formed from a given mass of reactant or product (Figure 2.2.10).

A known mass for use in a calculation to determine the unknown mass of an ionic compound is obtained by **gravimetric analysis**. In gravimetric analysis, the mass of a specific salt is determined by forming a precipitate with one of the soluble ions of the salt being investigated. The precipitate is then collected and weighed to determine its mass. Steps in a typical analysis are shown in Figure 2.2.11.

Table 2.2.6 Sources of salts in water

Source	Description	Examples
minerals	• naturally occurring solid salts in rocks that dissolve in water over time • run-off from artificial fertilisers, soaps and detergents	• calcium, magnesium, sodium, hydrogencarbonate, sulfate, chloride • sulfates, nitrates and phosphates
heavy metals	• metals with a high density that have a toxic effect on living things • can be from natural sources and/or industry	• lead, mercury, copper, arsenic
organometallic substances	• substances with at least one carbon-to-metal bond • used in laboratories and industry	• often catalysts or reagent in chemical processes, e.g. tetraethyl lead

 ISBN 978 0 6557 0015 9

KEY KNOWLEDGE

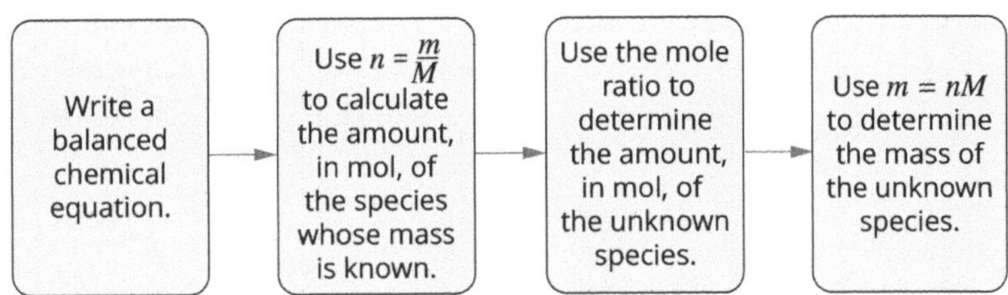

Figure 2.2.10 A flow chart for mass–mass stoichiometric calculations

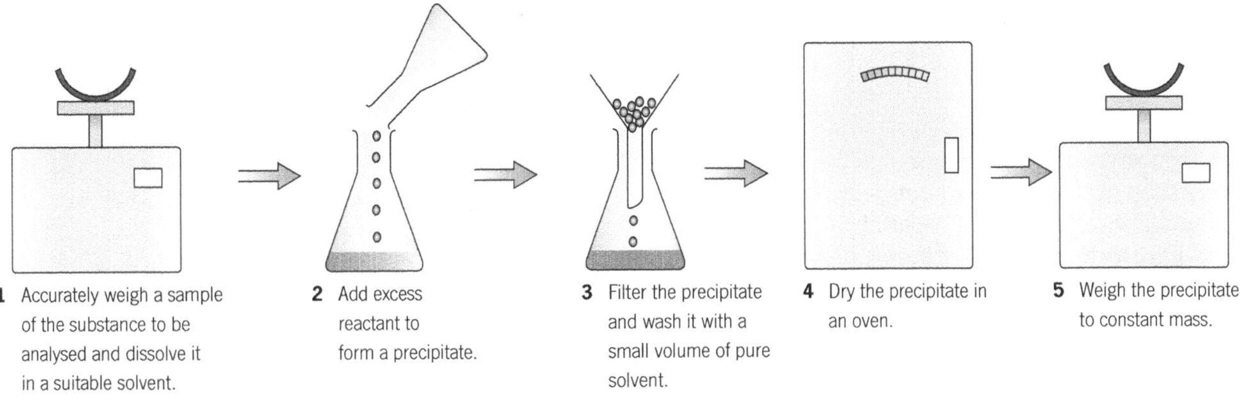

Figure 2.2.11 Steps in a gravimetric analysis

ANALYSING WATER BY COLORIMETRY OR UV-VISIBLE SPECTROSCOPY

Colorimetry and UV–visible spectroscopy are spectroscopic techniques that can be used to determine the concentration of ions or complexes in a water or soil sample.

The soluble salts that can be analysed by colorimetry or UV–visible spectroscopy include:

- iron(II) ions, Fe^{2+}, particularly if oxidised to iron(III) ions, Fe^{3+}, and mixed with potassium thiocynate to form the blood-red metal complex $FeSCN^{2+}$ (a metal complex consists of a central metal ion surrounded by an array of bound molecules of ions)
- cobalt ions also in the form of a metal complex
- phosphate ions, PO_4^{3-}.

The light energy selected for analysis of a particular component will be one that is absorbed strongly by the component but not absorbed by other components in the solution.

- When using colorimetry, the light chosen is the complementary colour (Figure 2.2.12) of the component being analysed.
- In UV–visible spectroscopy, a specific wavelength is used rather than a colour range, giving better sensitivity and precision than colorimetry.

The absorption of light energy causes electrons in ions or molecules to jump to higher energy levels. The amount of light energy absorbed increases as the concentration of the component increases.

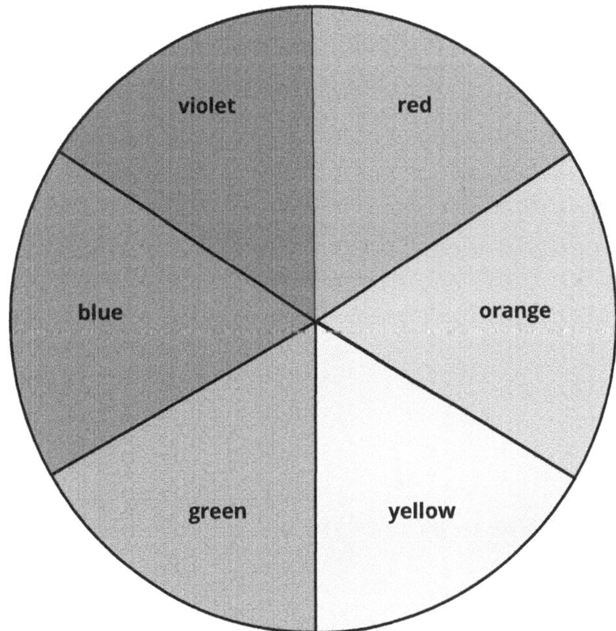

Figure 2.2.12 Complementary colours are opposite each other on a colour wheel.

ISBN 978 0 6557 0015 9

KEY KNOWLEDGE

The basic components of a UV–visible spectrometer are shown in Figure 2.2.13.

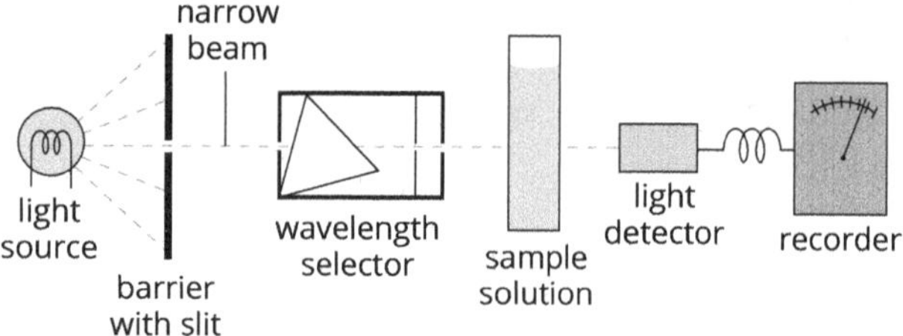

Figure 2.2.13 The basic components of a UV–visible spectrometer

Calibration curve

The concentration of a component in a sample is determined by using a calibration curve (Figure 2.2.14).

1 Standard solutions are prepared and a fixed volume of each is analysed. The absorption of light of the selected colour or wavelength by each standard solution is measured.
2 A calibration graph is plotted using the results obtained by the standard solutions.
3 The sample is analysed for its absorbance.
4 The sample's concentration is determined from its absorbance and the calibration curve.

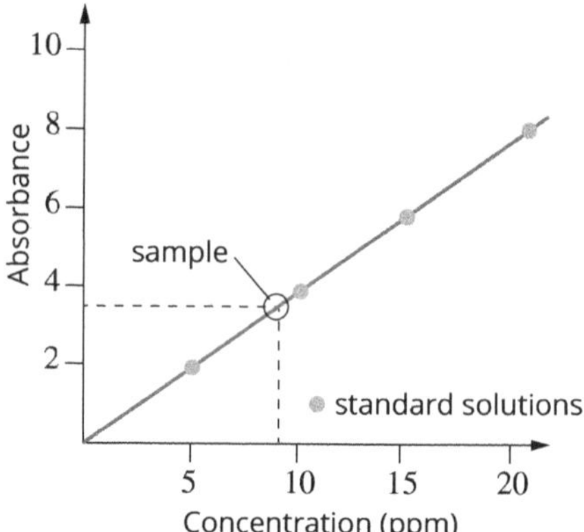

Figure 2.2.14 A calibration curve of concentration versus absorbance

- **You will now be able to complete Worksheets 41–43 and conduct Practical activities 26 and 27.**

 ISBN 978 0 6557 0015 9

Knowledge review—elements, compounds and molar mass

1 Mark each statement about elements, compounds, atoms and isotopes as true or false.

Statement	True or false?
An element contains only one or two different types of atom.	
A compound can exist as a molecule or lattice.	
A mixture contains elements and compounds that are not chemically bonded.	
$^{15}_{7}X$ and $^{15}_{8}Y$ are isotopes.	
The relative atomic mass of copper is 63.5.	
The ion $^{52}_{24}Cr^{2-}$ has 26 electrons.	
All elements in group 16 of the periodic table have 16 valence electrons.	
The standard mass to which other relative masses are compared is carbon-12, being exactly 10 units.	
The relative atomic mass is a weighted average of the isotopic masses of all of the isotopes that exist for an element.	

2 For each statement you marked as false, explain why it is incorrect.

3 Write the formula and molar mass, in g mol^{-1}, of each of the following compounds:

a magnesium chloride: formula: ______ molar mass: ______

b sodium carbonate: formula: ______ molar mass: ______

c aluminium oxide: formula: ______ molar mass: ______

d hydrochloric acid: formula: ______ molar mass: ______

e sodium nitrate: formula: ______ molar mass: ______

f sulfuric acid: formula: ______ molar mass: ______

4 A scientist wants to prepare a sample that contains 0.250 mol of sodium carbonate. What mass will the scientist need to measure?

ISBN 978 0 6557 0015 9

WORKSHEET 34

Molarity—measuring moles in solution

1 A number of similar terms are used when dealing with quantities in chemistry. Outline the use of each term below in the space provided.

Mole

Molar

Molar mass

Molarity

2 Each beaker below contains a solution of potassium chloride. Consider the diagram and answer the following questions.

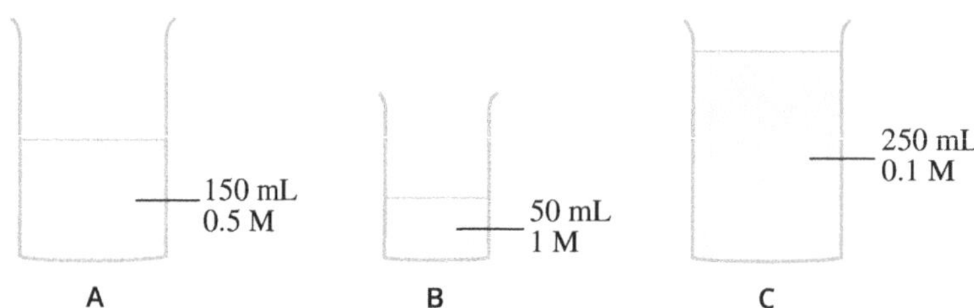

a Which beaker has the highest molarity—A, B or C?

b Calculate the amount, in moles, of potassium chloride dissolved in each beaker.

c Does the beaker with the highest molarity also have the greatest amount of dissolved solute? Explain your answer.

d A student adds water to beaker A until it holds the same volume of solution as beaker C. Calculate the new molarity in beaker A.

 ISBN 978 0 6557 0015 9

WORKSHEET 35

Converting between concentration units

1 A volumetric flask contains a solution in which 2.00 g of sodium chloride, NaCl, has been dissolved and made up to 250.0 mL.

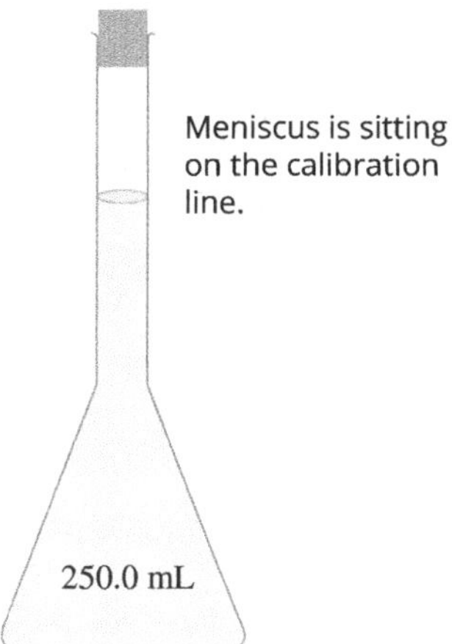

The concentration of this solution can be expressed using a number of different units. Calculate the concentration for each unit shown in the table below.

Unit	Concentration of the solution
$g\ L^{-1}$	
$mol\ L^{-1}$	
$mg\ L^{-1}$	
% (m/v)	
ppm	

2 For each solution in the following table, circle the concentrations that are equivalent to the concentration listed in the first column.

300 $mg\ L^{-1}$ NaCl solution	0.300 $g\ L^{-1}$	3.00 $g\ L^{-1}$	0.0051 M
25 ppm Mg^{2+} in water	25 $mg\ L^{-1}$	25 $g\ L^{-1}$	25 $\mu g\ g^{-1}$
12.5% (m/v) ethanol in wine	12.5 $g\ L^{-1}$	125 $g\ L^{-1}$	1.25×10^{6} ppm
2.0 M NaOH	0.050 $g\ L^{-1}$	$8.0 \times 10^{4}\ mg\ L^{-1}$	80 ppm

3 In your house, or on the internet, find a household product that is labelled with each of the concentration units listed in the table.

ISBN 978 0 6557 0015 9

WORKSHEET 36

Case study

Removing impurities from water using precipitation reactions

Read the following information about water hardness and then answer the questions below.

Water 'hardness' is caused by calcium (Ca^{2+}) and magnesium (Mg^{2+}) ions that have dissolved into water from minerals. The calcium and magnesium ions are impurities and although water hardness does not pose a danger to health, the dissolved ions can take part in reactions that leave insoluble mineral deposits such as scum rings on bathtubs and scale in kettles or industrial boilers. It is also harder to form a lather in hard water, so more soap or detergent is required in washing and in the shower. In Australia, water hardness is different in different states. Victoria and Tasmania have the softest water, whereas South Australia and Queensland have the hardest water.

To remove the impurities, hard water is treated by precipitation reactions, which involve calcium hydroxide ($Ca(OH)_2$) and sodium carbonate (Na_2CO_3). When calcium hydroxide is added to water, it dissolves to produce aqueous Ca^{2+} and OH^- ions, whereas sodium carbonate dissolves to produce Na^+ and CO_3^{2-} ions. Reactions then generate the insoluble precipitates calcium carbonate and magnesium hydroxide.

1 Define 'precipitation reaction'.

2 Write ionic equations for the precipitation reactions that produce the precipitates calcium carbonate and magnesium hydroxide.

3 Which of the dissolved ions is a spectator ion?

4 Explain why two compounds are needed for the removal of calcium and magnesium ions.

5 As well as the chemicals to cause precipitation, coagulants are also added. Carry out research to find out why coagulants need to be added and what their purpose is.

ISBN 978 0 6557 0015 9

WORKSHEET 37

Standard solutions

1 Tick the statements about standard solutions that are true.

- ☐ A standard solution has an accurately known concentration.
- ☐ A standard solution always has a neutral pH.
- ☐ A standard solution always has a concentration of 1.0 M.
- ☐ A standard solution can have any concentration, as long as it is accurately known.
- ☐ A standard solution can only be produced from a primary standard.

2 Equipment required for the preparation of a standard solution from a primary standard is shown below. Name each item.

A ______________________________

B ______________________________

C ______________________________

D ______________________________

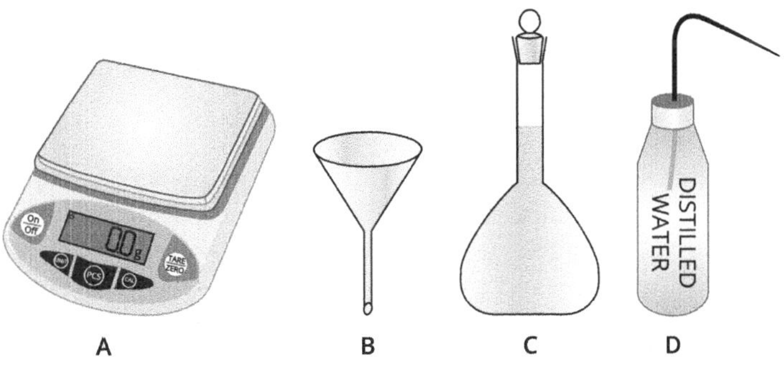

3 Determine the concentration of a standard solution prepared by dissolving 3.65 g of sodium carbonate, Na_2CO_3, in water and making the volume up to 250.0 mL in a volumetric flask.

____

____

____

____

____

4 Determine the mass of sodium carbonate you would need to dissolve in a 500 mL volumetric flask to make up 500 mL of a 1.00 M solution.

____

____

____

____

____

WORKSHEET 38

Acid-base titrations

Planning the steps of a volumetric analysis can help when you carry out the calculations. The formulas and prompts in the box may be helpful in your planning.

$n = cV$	calculate dilution factor	$c = \frac{n}{V}$	use mole ratio	write a balanced equation

For each acid–base titration calculation below:

- plan your method of solution by deciding which of the formulas you will use and in what order
- solve the problem, remembering to show your answer with the correct number of significant figures.

1 The ethanoic acid concentration, in M, in white vinegar was determined by titrating a 20.00 mL aliquot with 0.9925 M sodium hydroxide solution. The average titre volume was 20.75 mL of sodium hydroxide solution.

Plan:

Solution:

2 The ammonia content, in % (m/v), of cloudy ammonia was determined by first diluting a 20.00 mL sample to 250.0 mL in a volumetric flask. A 25.00 mL aliquot of this solution was titrated with 0.0975 M hydrochloric acid. The volume of acid used was 22.45 mL.

Plan:

Solution:

 ISBN 978 0 6557 0015 9

Mass–volume stoichiometry for gases

Stoichiometry allows calculation of the amounts of reactants consumed and products formed from a given amount of any one reactant or product. In mass–volume stoichiometry, the mass of a reactant or product can be used to determine the volume of a gaseous reactant or product, or the concentration of a solution. Stoichiometry always requires a balanced chemical equation to provide a mole ratio by which atoms react and products are formed.

A 10.0 g sample of magnesium carbonate is added to an excess amount of hydrochloric acid and allowed to completely react.

1 Write a balanced chemical equation for the reaction between $MgCO_3$ and HCl. Products of the chemical reaction are magnesium chloride, water and carbon dioxide gas.

2 Calculate the volume of carbon dioxide formed at 15°C and 120 kPa from the mass of magnesium carbonate reacted by completing the following steps.

a Calculate the amount, in mol, of $MgCO_3$.

b Identify the mole ratio of $MgCO_3$ reacted to CO_2 produced as given in the chemical equation.

c Use the mole ratio to calculate the amount, in mol, of CO_2 produced.

d Calculate the volume of CO_2 produced.

3 The other reactant in this reaction is hydrochloric acid. What is meant by describing it as 'in excess'?

4 Use a similar series of steps to those outlined in Question 2 to calculate the mass of magnesium carbonate required to react to produce 10.0 L of CO_2 at SLC.

WORKSHEET 40

Solving complex calculations—using more than one formula

Many calculations in chemistry require a number of different steps. Each of the following problems requires the use of at least two of the formulas shown.

$n = \frac{m}{M}$	$n = cV$	$n = \frac{N}{N_A}$
$pV = nRT$	$c_1V_1 = c_2V_2$	$n = \frac{V}{V_M}$

For each problem:

- plan your method of solution by deciding which of the formulas you will need to use and in what order
- solve the problem, remembering to show your answer with the correct number of significant figures.

The first problem has been started for you.

1 Calculate the mass, in grams, of solute dissolved in 700 mL of 0.10 M NaOH.

Plan: $n(\text{NaOH}) = cV$, then $m(\text{NaOH}) = nM$

Solve:

2 Calculate the number of HCl molecules present in 2.00 L HCl gas at 20°C and 150 kPa.

Solve:

3 Calculate the pH of a 150 mL aqueous solution in which 0.40 g NaOH is dissolved.

Plan:

Solve:

 ISBN 978 0 6557 0015 9

Analysis with light—colorimetry and UV-visible spectroscopy

1 Choose from the terms listed to complete the summary statements outlining the key points about colorimetry and UV–visible spectroscopy. Some terms are used more than once and not all terms are used.

concentration	mass	visible light	absorbance	ultraviolet and visible light
red light	blue light	absorbed	emitted	

- Spectroscopic techniques involve the measurement of the ______________ of a type of electromagnetic radiation by a sample.
- In colorimetry, the type of electromagnetic radiation is ______________. In UV–visible spectroscopy the type of electromagnetic radiation is ______________.
- In both colorimetry and UV–visible spectroscopy, the amount of light ______________ by a solution is directly related to the ______________ of the substance being analysed in the solution.

2 The concentration of iron(II) ions in a sample of water is to be analysed by UV–visible spectroscopy. The ions are first converted to the metal complex $FeSCN^{2+}$, which is a blood-red colour. The absorption spectrum of $FeSCN^{2+}$ is shown below.

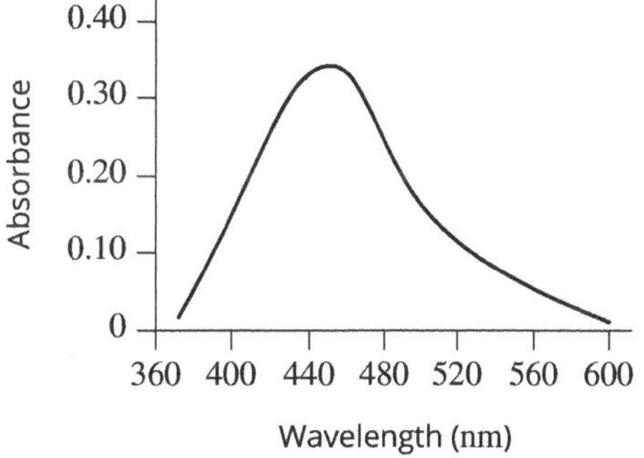

a i Which wavelength should be chosen for the analysis? Give a reason for your answer.

____

____

ii Give a reason why this wavelength might not actually be chosen.

____

____

b If the analysis were to be carried out by colorimetry and not UV–visible spectroscopy, what colour light would you choose for the analysis? Give a reason for your answer.

____

____

____

ISBN 978 0 6557 0015 9

3 A calibration curve for analysis of $FeSCN^{2+}$ is shown below.

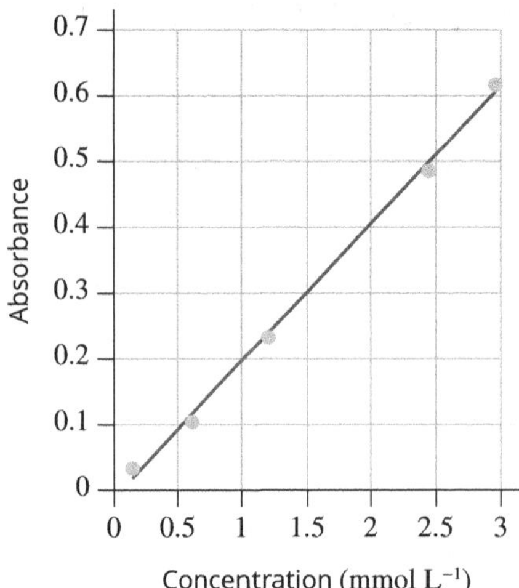

The absorbance of the sample of water is found to be 0.34.

a What is the concentration of Fe^{2+}, in mol L^{-1}?

b Calculate the concentration of Fe^{2+} ions in the water sample, in mg L^{-1}.

 ISBN 978 0 6557 0015 9

WORKSHEET 42

Literacy review—key terms and formulas

1 The table contains some descriptions of quantities measured in chemistry. Complete the table by writing the correct name of each quantity beside each description.

Description	Quantity
space occupied by a gas or solution	
force per unit surface area exerted by gas particles colliding with the walls of a container	
quantity of substance counted in lots of Avogadro's constant	
example of this quantity is 60 K	
mass of one mole of a substance	
volume of one mole of a gas	
amount of substance dissolved per litre of solution	

2 The table contains some formulas used in calculations of quantities. Write an explanation for each formula in the space beside each formula.

Formula	Explanation
$n = \frac{m}{M}$	
$c = \frac{n}{V}$	
$c_1V_1 = c_2V_2$	
$pV = nRT$	
$n = \frac{N}{N_A}$	
$V = n \times V_M$	

WORKSHEET 43

Reflection—How are chemicals measured and analysed?

The following table lists the key knowledge covered in this area of study.

1 Reflect on how well you understand the concepts listed. Rate your learning by shading the circle that corresponds to your current level of understanding for each one.

Key knowledge	Not confident ◄				► Very confident
Concentration—calculation using units of $mol\,L^{-1}$, $g\,L^{-1}$, %(m/v), %(v/v), ppm and conversion between units	○	○	○	○	○
Solubility graphs to predict ionic compound solubility	○	○	○	○	○
Analysis for acids and bases—titrations and volume–volume stoichiometry	○	○	○	○	○
Gases—major greenhouse gases and general properties of gases	○	○	○	○	○
Gas calculations—ideal gas equation, molar volume, mass–volume stoichiometry	○	○	○	○	○
Analysis for salts—electrical conductivity, mass–mass stoichiometry, mole ratio of water of hydration	○	○	○	○	○
Colorimetry and/or UV–visible spectroscopy—use of calibration curve to analyse ions or complexes	○	○	○	○	○

2 Consider the points you have shaded from 'Not confident' to 'Very confident'. List specific ideas you can identify that were challenging.

3 Write down two different strategies that you will apply to help further your understanding of these ideas.

 ISBN 978 0 6557 0015 9

PRACTICAL ACTIVITY 21

Controlled experiment

Determination of solubility of a salt in water

SUGGESTED DURATION

- 60 minutes

INTRODUCTION

The solubility of most ionic solids in water varies with temperature. As an unsaturated solution cools, it will reach a temperature at which crystallisation starts to occur. This indicates that the solution has become saturated. At this temperature, the solution contains the maximum quantity of solute that can be dissolved in that amount of solvent.

If the temperature at the point of saturation is recorded for several different quantities of solute, the data can be plotted on a graph in order to construct a solubility curve.

AIM

To investigate the solubility of potassium nitrate, KNO_3, in water at different temperatures and construct a solubility curve

MATERIALS

- 25 g potassium nitrate, KNO_3
- 20 mL deionised water
- electronic balance (2 decimal places)
- Bunsen burner, heat-resistant mat, tripod and wire gauze
- test tubes
- test-tube holder
- test-tube rack
- 500 mL glass beaker
- thermometer
- 10 mL measuring cylinder
- glass stirring rod
- retort stand and clamp
- spatula
- marking pen
- safety glasses/goggles
- disposable gloves

PRE-LAB SAFETY INFORMATION		
Material used/other risks	**Hazard**	**Control**
potassium nitrate	• may intensify fire; oxidiser • causes skin irritation • causes serious eye irritation • may cause respiratory irritation	Wear safety glasses/goggles, disposable gloves and a lab coat.

Please indicate that you have understood the information in the safety table.

Name (print): ____________________

I understand the safety information (signature): ____________________

METHOD

1. Put on your PPE, ensuring your lab coat is buttoned, your gloves are fitted correctly and comfortably, and your safety glasses/goggles are fitted over your eyes.
2. Use a marking pen to number four test tubes 1–4.
3. Measure exactly 2.0 g of potassium nitrate, KNO_3. Transfer it into test tube 1.
4. Repeat step 3 but adding 4.0 g, 6.0 g and 8.0 g of KNO_3 to test tubes 2–4.
5. Add exactly 5.0 mL of distilled water to each test tube.
6. Add 300 mL of tap water to a 500 mL glass beaker. This will be a water bath. Set up the water bath and test tube 1 as shown in the diagram. Heat the water to 90°C and adjust the Bunsen burner flame to maintain the water at this temperature.
7. Stir the solution in test tube 1 until the KNO_3 is completely dissolved. Remove the stirrer and rinse it in distilled water. Remove test tube 1 from the water bath.
8. Place a warm thermometer (dip it into the water bath first) into the solution in test tube 1. Hold the test tube up to the light and watch for the first sign of crystallisation in the solution. At the instant crystallisation starts, record the temperature in Table 1.
9. Repeat steps 7 and 8 for test tubes 2–4.
10. If any doubtful results are obtained, the procedure can be repeated by redissolving the salt in the hot water bath and allowing it to recrystallise.
11. Spray and wipe down your bench, return used equipment, wash equipment as necessary, and remove PPE as directed by your teacher.

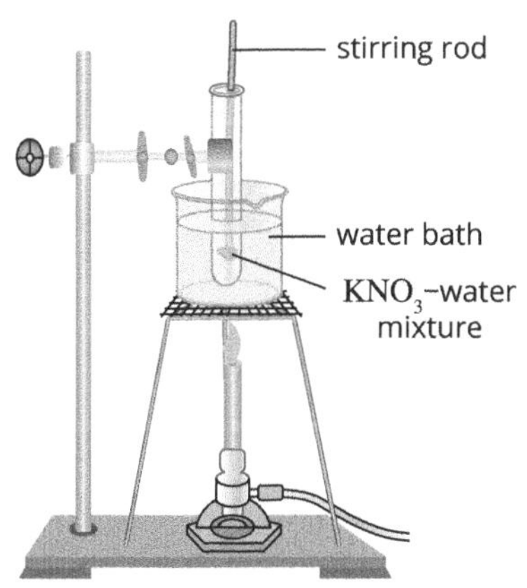

Experimental set-up

RESULTS

Table 1 Temperatures of crystallisation

Test tube	Mass (g) of KNO_3 in 5.0 mL of H_2O	Temperature at which crystallisation occurs (°C)
1	2.0	
2	4.0	
3	6.0	
4	8.0	

DISCUSSION

1 Convert the mass, in grams, in the 5.0 mL of water in the test tubes to solubility in g per 100 g of water. Record your answers in Table 2.

Table 2 Conversion of amounts dissolved to solubility

Test tube	Mass (g) of KNO_3 in 5.0 mL of H_2O	Solubility (g/100 g)
1	2.0	
2	4.0	
3	6.0	
4	8.0	

2 Plot your results on the grid below to create a solubility curve. Plot the solubility of KNO_3, in g/100 g, on the *y*-axis and the temperature on the *x*-axis.

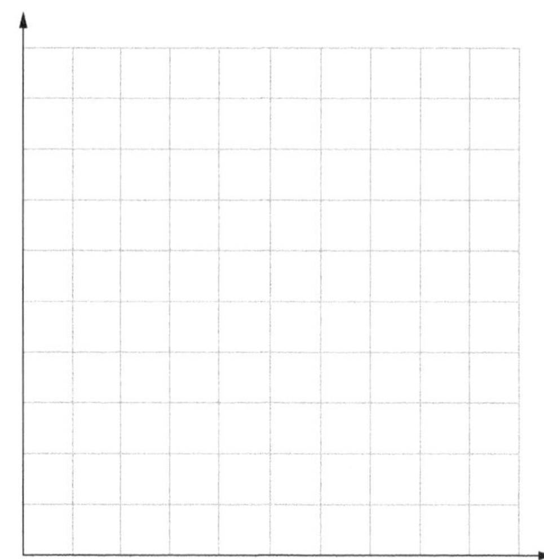

3 Use the solubility curve to determine the mass, in g, of KNO_3 that can be dissolved in 100 g of water at:

a 30°C ______

b 60°C ______

c 70°C ______

 ISBN 978 0 6557 0015 9

4 Classify the following solutions of KNO_3 as saturated, unsaturated or supersaturated. Explain your answers.

a 75 g/100 g at 40°C

b 60 g/100 g at 50°C

5 Evaluate whether the experiment provided you with an effective and efficient method of determining the solubility of a salt in water.

6 If you were to repeat the experiment, identify the steps you would do differently. In your answer, you should include how you:

- would change the method and how this might increase the accuracy and precision of the results
- performed the tasks and the skills that you need to improve on in your technique.

7 Explain how measurement of electrical conductivity could also be used as an indication of the amount of salt dissolved in a water sample.

CONCLUSION

PRACTICAL ACTIVITY 22

Controlled experiment

Preparation of a standard solution

SUGGESTED DURATION

- 30 minutes

INTRODUCTION

Standard solutions have accurately known concentrations and are used in titrations to determine the concentration of a solution of unknown concentration. They can be made by dissolving a measured mass of a primary standard in a known volume of solution.

AIM

To prepare a standard solution from a primary standard

PRE-LAB SAFETY INFORMATION		
Material used/other risks	**Hazard**	**Control**
anhydrous sodium carbonate	• causes serious eye irritation	Wear safety glasses/goggles.
Please indicate that you have understood the information in the safety table.		
Name (print):		
I understand the safety information (signature):		

MATERIALS

- anhydrous sodium carbonate, Na_2CO_3
- 250 mL deionised water
- 250 mL volumetric flask
- weighing bottle, weight boat or watch glass
- small funnel
- spatula
- plastic Pasteur pipette
- wash bottle containing deionised water
- electronic balance (3 or 4 decimal places if possible)
- paper label or marking pen
- safety glasses/goggles

PRE-LABORATORY EXERCISE

Calculate the mass of anhydrous sodium carbonate, Na_2CO_3, required to make 250 mL of 0.100 M solution.

METHOD

1. Put on your PPE, ensuring your safety glasses/goggles are fitted over your eyes.
2. Weigh out your calculated mass of anhydrous sodium carbonate (using a 3 decimal place or 4 decimal place balance if possible) in a weighing bottle or weight boat or on a watch glass. Record the exact mass of solid used in the Results section.
3. Use a small funnel to transfer the sodium carbonate, a little at a time, to a 250 mL volumetric flask. Use a wash bottle of deionised water to wash traces of sodium carbonate in the weighing bottle and funnel into the flask. Repeat the rinsing another two times. Be careful when using the wash bottle that you don't lose any reagent due to it splashing out while rinsing.
4. Add deionised water to the flask until it is almost half full. Stopper and swirl the contents of the flask to dissolve the sodium carbonate.
5. Add more water until the meniscus of the solution is nearly level with the calibration line. Use a dropping pipette to add the final few drops of deionised water so that the bottom of the meniscus is exactly level with the calibration line. Ensure the flask is standing on a horizontal surface and your eye is at the same level as the line. When using the dropper, try to avoid drops running down the neck of the flask as you may end up with a total volume of more than 250 mL.
6. Shake the flask by repeatedly inverting it and swirling it upside down so that the concentration of the solution is uniform. Label the flask with its contents and concentration and your name.
7. Spray and wipe down your bench, return used equipment, wash equipment as necessary, and remove PPE as directed by your teacher.

 ISBN 978 0 6557 0015 9

PRACTICAL ACTIVITY 22

RESULTS

Mass of anhydrous sodium carbonate: ________________ g

DISCUSSION

1 Calculate the concentration, in $mol\,L^{-1}$, of the sodium carbonate solution.

2 Convert the concentration you calculated in Question 1 to $g\,L^{-1}$ and ppm.

3 List the properties necessary for a chemical to be used as a primary standard.

4 Identify some of the sources of error associated with this experiment. Classify each error as systematic or random.

5 How would the accuracy of the concentration of the standard solution be affected if the volumetric flask had been rinsed with deionised water before use and droplets of water were left in the flask when the sodium carbonate was added?

CONCLUSION

PRACTICAL ACTIVITY 23

Controlled experiment

Determination of HCl content in brick cleaner

SUGGESTED DURATION

- 60 minutes

INTRODUCTION

Brick cleaner contains concentrated hydrochloric acid as the active ingredient. The acid reacts with the basic components of concrete, enabling concrete to be removed from brickwork. To analyse brick cleaner, a sample is diluted (because the original acid is highly concentrated) and titrated against a standard solution of a base, sodium carbonate.

AIM

To use the standard solution prepared in Practical activity 22 to find the percentage, by mass, of hydrogen chloride (present as hydrochloric acid) in brick cleaner using an acid–base titration

As directed by your teacher, complete the pre-lab safety information by referring to the safety data sheets (SDSs) or your teacher's risk assessment for the activity.

PRE-LAB SAFETY INFORMATION

Material used/other risks	Hazard	Control
brick cleaner (concentrated hydrochloric acid, ~30% w/w)		
0.1 M sodium carbonate solution		
methyl orange		

Please indicate that you have understood the information in the safety table.

Name (print):

I understand the safety information (signature):

MATERIALS

- 5 mL brick cleaner or concentrated hydrochloric acid, HCl
- 100 mL of approximately 0.1 M standard sodium carbonate solution, Na_2CO_3 (The standard solution prepared in Practical activity 22 can be used.)
- 250 mL deionised water
- methyl orange indicator
- dropper bottle of 250 mL volumetric flask
- 100 mL glass beaker
- 4 × 100 mL glass conical flasks
- 2 small funnels
- 10 mL glass measuring cylinder
- 20 mL bulb pipette
- pipette filler
- plastic Pasteur pipette
- burette and stand with burette clamp
- white tile
- electronic balance (3 decimal places)
- safety glasses/goggles
- disposable gloves

METHOD

1. Put on your PPE according to the risk assessment and management table you filled in. As relevant, ensure your lab coat is buttoned, your gloves are fitted correctly and comfortably, and your safety glasses/goggles are fitted over your eyes.
2. Weigh a clean, dry 100 mL beaker and record its mass in Table 1.
3. Note the hydrogen chloride content of the brick cleaner as specified by the manufacturer. In a fume hood, use a 10 mL measuring cylinder to measure about 5 mL of brick cleaner. Pour this into the 100 mL beaker, avoiding spillages.
4. Reweigh the beaker plus contents and record the mass in Table 1.
5. Add deionised water until the volumetric flask is about half full. Quantitatively transfer, using a funnel, the brick cleaner/acid from the beaker into flask. Carefully add a small amount of water with a wash bottle, after the initial transfer, to the beaker (sit it on the bench, do not hold it at this stage), adding a little more after 15 seconds. Gently swirl it around, and transfer it to the flask. Repeat two more times, ensuring you rinse down the sides of the beaker. Stopper the flask and turn it upside down carefully several times to mix the solution thoroughly.
6. Remove the flask from the fume hood, and return it to your bench. Add more water to the flask until the meniscus is level with the calibration line. Stopper the flask and mix the solution thoroughly.
7. Use a 20 mL pipette to place 20.00 mL aliquots of standard sodium carbonate solution into four 100 mL conical flasks. Add two to three drops of methyl orange indicator to each conical flask. Set one flask aside as a control in colour matching.

 ISBN 978 0 6557 0015 9

8 ▪ Fill a burette with the diluted solution of brick cleaner. Wait a few minutes until no more air bubbles rise to the surface. Record the initial level of the solution in the burette to two decimal places. Remember, you can only record measurements on quantitative glassware to a resolution of halfway between their smallest measurement, i.e. 1.05 mL.

9 ▪ Titrate the sodium carbonate solution with the dilute solution of brick cleaner until the moment when the indicator just shows a permanent colour change from yellow to orange. Record the final burette reading in Table 2 and calculate the volume of the titre. This colour change can be subtle, so slow down to drop by drop when you think you are close. If you are unsure if it needs 'one drop' more, record the volume in the burette before adding one more drop, in case it was already at end point.

10 ▪ Repeat the titration until you have three concordant titres (i.e. the smallest is no more than 0.1 mL less than the largest).

11 ▪ Spray and wipe down your bench, return used equipment, wash equipment as necessary, and remove PPE as directed by your teacher.

RESULTS

Table 1 Brand, mass and concentration results

Brand of concrete cleaner	
Hydrogen chloride content specified by manufacturer	
Mass of 100 mL glass beaker (g)	
Mass of 100 mL glass beaker and brick cleaner (g)	
Concentration of standard Na_2CO_3 solution (M)	

Table 2 Titration results

Titration	Initial burette reading	Final burette reading	Titre (mL)
1			
2			
3			
4			

DISCUSSION

1 Determine the mass of the sample of brick cleaner.

2 Find the measurement result of titre volume to use in your calculations.

3 Calculate the amount of sodium carbonate, in mol, present in each conical flask.

4 Write a balanced equation for the reaction that occurs between hydrochloric acid and sodium carbonate solution.

5 Calculate the amount of hydrochloric acid, in mol, present in the average titre.

6 Calculate the amount of hydrochloric acid, in mol, present in the volumetric flask.

7 Find the percentage, by mass, of HCl in the brick cleaner.

8 a How does your answer to Question 7 compare with the manufacturer's claim about the composition of the brick cleaner?

b Compare your result with those of other members in your class. Explain why differences arise and how a more accurate result could be obtained.

9 Identify some sources of systematic error that could have affected the accuracy of your result.

10 Suppose you wished to rinse, but not dry, the following apparatus before use in this experiment. Name the liquids that should be used in each case and briefly explain why the liquids are chosen.

Apparatus	Rinse with	Explanation
a Volumetric flask		
b Burette		
c Pipette		
d Conical flask		

CONCLUSION

 ISBN 978 0 6557 0015 9

PRACTICAL ACTIVITY 24

Controlled experiment

Investigating the volume-pressure relationship in gases

SUGGESTED DURATION

- 30 minutes

INTRODUCTION

The ideal gas law links the gas properties of volume, pressure, amount and temperature. If two of these quantities are held constant, such as temperature and amount, then the relationship between the remaining two quantities can be investigated.

AIM

To develop a quantitative relationship between gas pressure and volume for a fixed amount of gas at constant temperature

MATERIALS

- retort stand, clamp and bosshead
- 50 mL disposable plastic syringe
- rubber stopper with holes bored part-way through
- 5 or 6 weights of equal mass (e.g. 500 g weights or half bricks weighing approx. 2 kg)

PRE-LAB SAFETY INFORMATION

Material used/other risks	Hazard	Control
heavy weights	• injury or damage due to falling weights	Conduct investigation on a tray and/or the floor (if clean).

Please indicate that you have understood the information in the safety table.

Name (print): ______________________

I understand the safety information (signature): ______________________

METHOD

1. Clamp the syringe onto the retort stand with the nozzle downwards and the plunger set at full capacity. Fit the rubber stopper to the nozzle and record the volume of air in the syringe at full capacity.
2. Place a 500 g weight or half brick on top of the plunger and record the volume of air in the syringe. Check your result by repeating this step twice. Record each reading and calculate the average volume.
3. Place a second weight or half brick on top of the first and record the volume. Check and average your results as in step 2.
4. Repeat this procedure until a total of five or six weights have been used.
5. Calculate and record the reciprocal of each average volume.

RESULTS

Table 1 Volume results

Weight (kg)	Volume (mL)				Reciprocal of volume $\left(\frac{1}{V}\right)$
	Trial 1	Trial 2	Trial 3	Average volume (mL)	

DISCUSSION

1 Produce a graph by plotting average volume on the *y*-axis against pressure (represented by the mass of weights or bricks) on the *x*-axis. This is graph A. Join the points with a line of best fit.

2 Plot the reciprocal of volume on the *y*-axis against pressure on the *x*-axis. This is graph B. Join the points with a line of best fit. To allow for the pressure that is caused by the atmosphere and by the mass of the plunger, raise the *x*-axis of graph B until the line passes through the origin.

3 Use your observations to write a description of how the volume of a fixed amount of air at a fixed temperature varies with pressure.

4 What type of mathematical relationship between the volume of a gas and its pressure does graph B suggest?

 ISBN 978 0 6557 0015 9

5 Explain the relationship you have identified in Questions 3 and 4.

CONCLUSION

PRACTICAL ACTIVITY 25

Determining the molar volume of hydrogen

SUGGESTED DURATION

- 50 minutes

INTRODUCTION

In this experiment, you will measure the volume of hydrogen gas produced in a reaction between magnesium metal and hydrochloric acid. The mass of magnesium will be used to determine the amount, in mol, of hydrogen gas produced. Along with further measurements of gas pressure and temperaure, the molar volume of hydrogen at standard laboratory conditions (SLC) can be calculated.

AIM

To determine the molar volume of hydrogen gas at SLC, 25°C and 100 kPa

PRE-LAB SAFETY INFORMATION		
Material used/other risks	**Hazard**	**Control**
2 M HCl	• non-hazardous < 3 M	Wear safety glasses/ goggles.

Please indicate that you have understood the information in the safety table.

Name (print): ______________________

I understand the safety information (signature): ______________________

MATERIALS

- 20 mL of 2 M HCl
- 8 cm length of magnesium ribbon, Mg, that has a mass of no more than 0.08 g
- 100 mL gas syringe
- set of apparatus to clamp the syringe to a retort stand
- 125 mL conical flask
- one-hole stopper to fit conical flask
- retort stand, bosshead and clamp to support conical flask
- 4 cm length of glass tubing to fit the one-hole stopper
- approx. 50 cm length of rubber tubing to connect the gas syringe to the glass tubing through the one-hole stopper
- electronic balance (2 decimal places)
- emery paper (fabric-backed sandpaper) or steel wool for cleaning the magnesium ribbon
- safety glasses/goggles

As directed by your teacher, complete the pre-lab safety information by referring to the safety data sheets (SDSs) or your teacher's risk assessment for the activity. Take note of the safety requirements for hydrogen produced in small quantities.

METHOD

1. Put on your PPE, ensuring your safety glasses/goggles are fitted over your eyes.
2. Clamp the stoppered test tube and gas syringe to their respective retort stands and connect the conical flask and syringe using the rubber tubing as shown in the diagram below. Check that the equipment is secure.

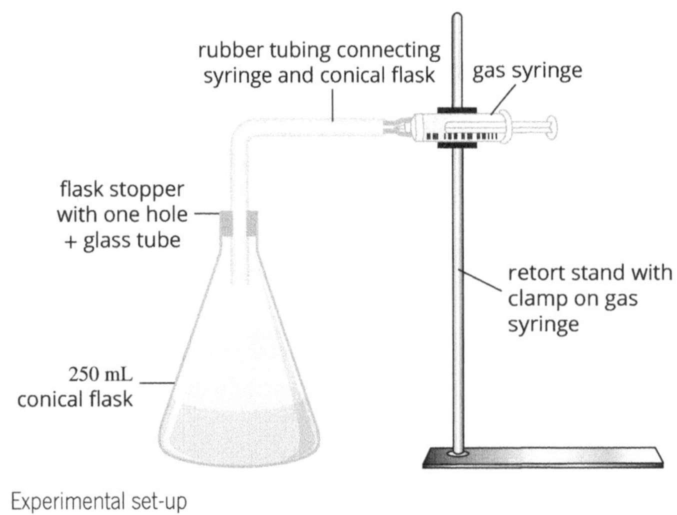

Experimental set-up

3. Remove the stopper from the conical flask and carefully pour about 15 mL of 2 M hydrochloric acid into the conical flask down one side, keeping the other side dry.
4. Clean and accurately weigh the magnesium ribbon, making sure that it weighs no more than 0.08 g.

 ISBN 978 0 6557 0015 9

5 ▪ Tilt the conical flask and carefully place the magnesium ribbon on the dry side of the conical flask, making sure that the magnesium does not contact the acid. Replace the stopper tightly.

6 ▪ Carefully withdraw the plunger of the syringe and then release it. If the system has no leaks, the plunger will return to its original position. Once any leaks have been fixed, record the initial volume shown on the syringe in Table 1.

7 ▪ Tilt or shake the conical flask so that the magnesium contacts the acid. As gas fills the syringe, rotate the plunger gently to prevent it from sticking.

8 ▪ Once the magnesium has been used up, allow the conical flask to cool. Record the final volume of gas in the syringe when the plunger has completely stopped moving. Calculate and record the increase in the volume of gas in the syringe.

9 ▪ Record the room temperature and atmospheric pressure.

10 ▪ From data supplied by your teacher, record the water vapour pressure at room temperature. To allow for the pressure of water vapour in the syringe, subtract the water vapour pressure from atmospheric pressure.

11 ▪ Spray and wipe down your bench, return used equipment, wash equipment as necessary, and remove PPE as directed by your teacher.

RESULTS

Table 1 Results

mass of magnesium (g)	
initial volume of syringe (mL)	
final volume of syringe (mL)	
volume of gas (mL)	
room temperature (K)	
atmospheric pressure p_{atm} (kPa)	
water vapour pressure at measured temperature $p(H_2O)$ (kPa)	
pressure of H_2 gas = $p_{atm} - p_{vapour}$ (kPa)	

DISCUSSION

1 Write a balanced equation for the reaction between magnesium and hydrochloric acid.

2 Calculate the amount, in moles, of magnesium reacted.

3 Using the amount of magnesium reacted, calculate the amount, in moles, of H_2 gas produced.

4 This amount of H_2 gas occupied the measured volume at the experimental pressure and temperature. Calculate the volume occupied by one mole (molar volume) at these experimental conditions.

5 Convert your answer to Question 4 to molar volume at standard laboratory conditions (SLC), 25°C and 100 kPa.

6 How close is your result for the molar volume of hydrogen to the accepted molar volume of an ideal gas? Suggest two sources of error for the molar volume you have calculated.

7 Explain why the molar volume of any gas at specified conditions, such as SLC, is the same.

8 Using your measurements for the volume, temperature and pressure of the hydrogen gas sample and the ideal gas equation, calculate the amount, in moles, of H_2 gas produced. Compare your answer with your answer to Question 3. How well do the two answers agree?

CONCLUSION

 ISBN 978 0 6557 0015 9

PRACTICAL ACTIVITY 26

Controlled experiment

Gravimetric determination of sulfur as sulfate in fertiliser

SUGGESTED DURATION

- 90 minutes over 2 or 3 days

INTRODUCTION

Sulfur is an essential mineral for plants because it is needed for the synthesis of chlorophyll and for nitrogen fixation in legumes. Therefore, the sulfur content of fertilisers is an important consideration for farmers and home gardeners.

This practical activity determines the amount of sulfur in a fertiliser in the form of sulfate ions. The sulfate is precipitated as barium sulfate from a solution containing a known mass of the fertiliser by adding an excess of barium chloride solution:

$Ba^{2+}(aq) + SO_4^{2-}(aq) \rightarrow BaSO_4(s)$

The proportion of sulfate ions and therefore of sulfur in the fertiliser is determined by collecting and weighing the precipitate that is formed during the above reaction and the use of mass–mass stoichiometry.

AIM

To find the percentage of sulfur (as sulfate) present in a brand of commercial fertiliser

As directed by your teacher, complete the pre-lab safety information by referring to the hazard labels on the reagent bottles or safety data sheets (SDSs) or your teacher's risk assessment for the activity.

PRE-LAB SAFETY INFORMATION		
Material used/other risks	**Hazard**	**Control**
fertiliser	• can cause skin, eye and lung irritation	Wear safety glasses/ goggles, disposable gloves and a lab coat.
0.5 M barium chloride solution		
0.1 M silver nitrate solution		
2 M hydrochloric acid		
heating of solution and crucible		

Please indicate that you have understood the information in the safety table.

Name (print):

I understand the safety information (signature):

MATERIALS

- 1.0 g fertiliser, finely ground using a mortar and pestle
- 3 mL of 2 M hydrochloric acid, HCl
- 5 mL of 0.1 M silver nitrate solution, $AgNO_3$
- 20 mL of 0.5 M barium chloride solution, $BaCl_2$
- 200 mL deionised water
- 50 mL warm deionised water
- 2 × 100 mL glass beakers
- 600 mL glass beaker
- glass filter funnel
- filter paper
- 10 mL glass measuring cylinder
- 25 mL glass measuring cylinder
- 50 mL glass measuring cylinder
- burette and stand with burette clamp
- vacuum flask and vacuum pump (water-jet type)
- glass filter crucible (No. 4 porosity) and rubber adaptor
- glass stirring rod
- wash bottle containing deionised water
- mortar and pestle
- Bunsen burner, tripod stand and gauze mat
- bench mat
- electronic balance
- oven
- safety glasses/goggles
- disposable gloves

PRACTICAL ACTIVITY 26

METHOD

1 ▪ Put on your PPE according to the risk assessment and management table you filled in. As relevant, ensure your lab coat is buttoned, your gloves are fitted correctly and comfortably, and your safety glasses/goggles are fitted over your eyes.

2 ▪ Accurately weigh out 1.0 g of the finely ground fertiliser into a preweighed 100 mL glass beaker. Record the mass in Table 1 and, if a commercial fertiliser is being analysed, record the brand and its sulfur content as specified by the manufacturer.

3 ▪ Add 50 mL of deionised water and stir to dissolve as much of the sample as possible. Filter the mixture, using a funnel and filter paper (gravity flow), into a 600 mL beaker, washing the residue (the solid left in the filter paper) several times with deionised water. If desired, the mixture can be left to stand overnight at this stage.

4 ▪ Add about 3 mL of 2 M hydrochloric acid to the filtrate and add more water so that the total volume is about 200 mL. Heat the solution until it boils.

5 ▪ Add 15 mL of 0.5 M barium chloride solution drop by drop from a burette to the hot solution, stirring continuously. A white precipitate of barium sulfate will form.

6 ▪ Boil the mixture for a further minute, then remove it from the heat and allow the precipitate to settle (10–15 minutes). Ensure no sulfate ions remain in the solution by adding several more drops of barium choride solution. If more precipitate forms, add 3 mL of barium chloride solution and test again for unreacted sulfate ions. If desired, the mixture can be left to stand overnight at this stage.

7 ▪ Weigh a glass filter crucible and record the mass.

8 ▪ Collect the precipitate in the glass filter crucible, using gentle vacuum filtration. (Filtration is faster if most of the liquid is filtered before the bulk of the precipitate is collected in the crucible.) Use about 10 mL of warm deionised water to wash any precipitate remaining in the beaker into the crucible. If using a vacuum pressurised Installed system, use only gentle suction force or you may tear the filter paper.

9 ▪ Collect the last drops of filtrate in a 100 mL beaker and test for the presence of chloride ions by adding a few drops of silver nitrate solution to the filtrate. If the solution becomes cloudy, wash the precipitate with a further 10 mL of warm water and repeat the test.

10 ▪ Place the crucible and contents in an oven heated to 100–110°C and leave overnight.

11 ▪ Weigh the crucible and contents and record the mass in Table 1.

12 ▪ Spray and wipe down your bench, return used equipment, wash equipment as necessary, and remove PPE as directed by your teacher.

RESULTS

Table 1 Brand and mass results	
Brand of fertiliser	
Mass of fertiliser sample (g)	
Sulfur as sulfate (content specified by manufacturer)	
Mass of filter crucible (g)	
Mass of filter crucible and contents (g)	

DISCUSSION

1 Determine the mass of the barium sulfate precipitate obtained.

2 Calculate the mass of sulfate ions present in the barium sulfate precipitate.

ISBN 978 0 6557 0015 9

3 Determine the percentage by mass of sulfate in the fertiliser.

4 Determine the percentage by mass of sulfur (as sulfate) in the fertiliser.

5 How could the results obtained be affected if the:

a mixture in step 6 of the method was not tested with more barium chloride solution?

b filtrate was not tested with silver nitrate solution when the precipitate was washed in step 9?

6 What sources of systematic and random error might arise in a gravimetric analysis such as this? Comment on the accuracy and precision of your results.

7 Compare your result to results obtained by other groups. Comment on the reproducibility of your results.

8 Agricultural fertilisers are one source of salts found in water or soil in the environment. Identify two other sources of salts.

CONCLUSION

PRACTICAL ACTIVITY 27

Controlled experiment

Colorimetric determination of phosphorus content of lawn fertiliser

SUGGESTED DURATION

- 50 minutes

INTRODUCTION

When a solution containing phosphate ions is added to a solution of ammonium molybdate reagent, an intense blue compound is formed. Ammonium molybdate converts phosphates to phosphomolybdate. Phosphomolybdate is then reduced by addition of ascorbic acid. The intensity of colour of the blue compound that develops is dependent on the concentration of the phosphate in the original solution. The reaction is not instantaneous. The blue reaches its most intense shade after about 10 minutes, after which it gradually fades. By comparing the intensity of the colour of a sample solution that has an unknown phosphate concentration with the colours of a series of standard solutions, the concentration of an unknown solution can be estimated. This analytical technique is known as colorimetry

AIM

To determine the phosphorus content of a lawn fertiliser by colorimetric analysis

PRE-LAB SAFETY INFORMATION		
Material used/other risks	**Hazard**	**Control**
ammonium molybdate		
10 mg L^{-1} phosphate solution		
ascorbic acid		

Please indicate that you have understood the information in the safety table.

Name (print): ________________

I understand the safety information (signature): ________________

MATERIALS

- fertiliser powder, finely ground
- 12 mL ammonium molybdate reagent
- ascorbic acid (do not use calcium salt)
- 50 mL 10.0 mg L^{-1} standard phosphate solution
- deionised water
- 20 mL hot deionised water
- 2 × 10 mL measuring cylinders
- 2 × 25 mL glass measuring cylinders
- 2 × 250 mL glass measuring cylinders
- 100 mL glass beaker
- 2 × 600 mL glass beakers
- 6 large test tubes
- test-tube rack
- test-tube holder
- glass stirring rod
- marking pen
- Bunsen burner and bench mat
- electronic balance (3 decimal places)
- colorimeter and data collection system
- safety glasses/goggles
- heat gloves

As directed by your teacher, complete the pre-lab safety information by referring to the hazard labels on the reagent bottles or safety data sheets (SDSs) or your teacher's risk assessment for the activity.

METHOD

Put on your PPE, ensuring your lab coat is buttoned, your gloves are fitted correctly and comfortably, and your safety glasses/goggles are fitted over your eyes.

PART A • PREPARATION OF SOLUTIONS

If the fertiliser contains about 1.3% phosphorus in the form of water-soluble phosphate, calculate the approximate mass of fertiliser required to prepare the 250 mL solution.

____

____

 ISBN 978 0 6557 0015 9

1 ▪ If a commercial fertiliser is to be analysed, record the brand and its phosphorus content (as water-soluble phosphate) as specified by the manufacturer. Accurately weigh 0.0310 g of finely ground fertiliser into a 100 mL beaker and record the mass.

2 ▪ Add about 20 mL hot water and stir to dissolve the powder.

3 ▪ Transfer the mixture to a 250 mL measuring cylinder, using water to rinse into the cylinder traces of fertiliser solution left in the beaker. Add water to the cylinder to make 250 mL of solution. Transfer the solution to a 600 mL beaker and stir to ensure the concentration is uniform.

4 ▪ Dilute the fertiliser solution by a factor of 10 by pouring 25 mL of the fertiliser solution into a clean 250 mL measuring cylinder and adding water to make 250 mL of solution. Mix well. Discard the remainder of the fertiliser solution.

5 ▪ Label six test tubes as follows: '10.0 mg L^{-1}', '7.5 mg L^{-1}', '5.0 mg L^{-1}', '2.5 mg L^{-1}', '0.0 mg L^{-1}' and 'unknown'.

6 ▪ Place 20 mL of the diluted fertiliser solution prepared in step 4, using a clean 25 mL measuring cylinder, into the test tube labelled 'unknown'.

7 ▪ Use a 10 mL measuring cylinder to add standard phosphate solution and water to the labelled test tubes according to Table 1.

Table 1 Phosphate solution standard dilutions

Test-tube label (mg L^{-1})	Volume of 10 mg L^{-1} standard phosphate solution (mL)	Volume of water (mL)
10.0	20	0
7.5	15	5
5.0	10	10
2.5	5	15
0.0	0	20

8 ▪ Add 2 mL ammonium molybdate reagent and a few crystals of ascorbic acid to each of the six test tubes. Stir each test tube to dissolve the crystals.

9 ▪ Place the six test tubes in a 600 mL beaker containing 200 mL boiling water. Heat for 5 minutes. On heating, the solutions should turn blue. Remove all test tubes at the same time and allow them to cool (10–15 minutes). (Heating time affects the intensity of colour obtained, so it is important that each tube be heated in exactly the same way for the same length of time.)

10 ▪ Compare the colour of the solution in the test tube labelled 'unknown' with the colour of the standards to estimate the concentration of phosphate in the diluted fertiliser solution.

PART B • USE OF COLORIMETER AND ELECTRONIC DATA COLLECTION EQUIPMENT

1 ▪ Fill a cuvette to three-quarters of its volume with the 0 mg L^{-1} solution and wipe the outside of the cell with a tissue. By following the manufacturer's instructions, calibrate the colorimeter to read zero transmittance when no light passes through the cell and 100% transmittance when yellow light (565 nm) passes through the cuvette. Use yellow light for the remainder of this experiment.

2 ▪ Discard the liquid from the cuvette, rinse the cuvette twice with the 2.5 mg L^{-1} solution and fill the cuvette to three-quarters of its volume with this solution. Measure the absorbance and record the results in Table 2. In a similar fashion, measure the absorbance of the other standard solutions.

3 ▪ Measure the absorbance of the solution of unknown concentration and determine, using the graph, the concentration of phosphate ions in this sample.

4 ▪ Spray and wipe down your bench, return used equipment, wash equipment as necessary, and remove PPE as directed by your teacher

RESULTS

PART A

Estimate of concentration of phosphate in diluted fertiliser: ______________________

PART B

Table 2 Absorbance of standard solutions	
Sample (mg L^{-1})	**Absorbance**
10.0	
7.5	
5.0	
2.5	
0.0	
diluted fertiliser sample	

DISCUSSION

1 Construct a graph of absorbance against concentration (using data collection or spreadsheet software if possible). Because absorbance is directly proportional to concentration, you can draw a straight line of best fit through the data points and passing through the origin.

2 What is the concentration of phosphate ions in the diluted sample?

3 The initial 250 mL of fertiliser solution was diluted by a factor of 10 before being analysed. What is the concentration of phosphorus ions, in mg L^{-1}, in the initial 250 mL of fertiliser solution?

4 Use your answer to Question 3 to find the mass of phosphate ions in the initial 250 mL solution. This is the same as the mass of phosphate in the sample of fertiliser.

5 Calculate the percentage of phosphorus, by mass, in the fertiliser sample.

 ISBN 978 0 6557 0015 9

6 If you analysed a commercial fertiliser, how well does your result for the percentage of phosphorus present agree with the manufacturer's specification? How does it compare with the estimate you made in Part A? Try to account for any difference between your result and the manufacturer's specification by referring to possible errors.

7 Why was yellow light used to measure the absorbance of solutions rather than blue light?

CONCLUSION

EXAM-STYLE QUESTIONS

Multiple-choice questions

Question 1

Which of the following solutions of sodium hydroxide, NaOH, is the most concentrated?

A. $1.0\,mol\,L^{-1}$

B. $30\,g\,L^{-1}$

C. 500 ppm

D. $2000\,mg\,L^{-1}$

Question 2

Identify the species whose solubility in water will decrease as temperature increases from room temperature.

A. CO_2

B. KNO_3

C. Na_2CO_3

D. $PbSO_4$

Question 3

Which of the following ions could be completely removed from water by a precipitation reaction with hydroxide ions?

A. Na^+

B. K^+

C. Mg^{2+}

D. Ca^{2+}

Question 4

What is the correct definition of a standard solution?

A. a solution that can only be produced from a primary standard

B. a solution that has an accurately known concentration

C. a solution that has a concentration of 1.0 M

D. a solution that is used in volumetric analysis

Question 5

Select the best description of the end point in an acid–base titration.

A. the point at which the pH of the solution is equal to 7

B. the point at which equal numbers of moles of acid and base have been mixed

C. the point at which the indicator changes colour

D. the point at which the reactants have been mixed in an equal stoichiometric ratio

Question 6

What is the volume of 20.0 g of propane gas, C_3H_8, at 80.0 kPa and 5°C?

A. 0.236 L

B. 13.1 L

C. 63.5 L

D. 578 L

ISBN 978 0 6557 0015 9

EXAM-STYLE QUESTIONS

Question 7

Which of the following gases will occupy the largest volume at SLC (standard laboratory conditions)?

A. 44.0 g CO_2

B. 32.0 g O_2

C. 58.0 g C_4H_{10}

D. They all have the same volume.

Question 8

Which one of the following is always used in a gravimetric analysis that includes mass–mass stoichiometry?

A. indicator

B. burette

C. volumetric flask

D. balance

Question 9

Consider the following chemical equation:

$Al_2O_3(s) + 6HCl(aq) \rightarrow 2AlCl_3(aq) + 3H_2O(l)$

When 2 mol of Al_2O_3 completely reacts in excess hydrochloric acid, what will be the amount, in mol, of $AlCl_3$ produced?

A. 1 mol

B. 2 mol

C. 4 mol

D. 6 mol

Question 10

Which of the following solutions would be expected to have the highest electrical conductivity?

A. 1.0 M NaCl

B. 2.0 M NaCl

C. 1.0 M Na_3N

D. 2.0 M Na_3N

Short-answer questions

Question 1

A student wishes to prepare 500 mL of a 0.100 M standard solution of sodium carbonate, Na_2CO_3.

a. Calculate the mass of sodium carbonate required. 2 marks

b. Write a method for the student to follow. 3 marks

EXAM-STYLE QUESTIONS

Question 2

Calculate the mass, in kg, of carbon dioxide gas produced from the combustion of 1.00 kg of butane gas according to the following equation. 3 marks

$2C_4H_{10}(g) + 13O_2(g) \rightarrow 8CO_2(g) + 10H_2O(g)$

Question 3

Calculate the volume, in mL, of 2.00 M nitric acid, at SLC, required to completely react with 3.45 g of magnesium metal according to the following equation. 3 marks

$Mg(s) + 2HNO_3(aq) \rightarrow Mg(NO_3)_2(aq) + H_2(g)$

Question 4

Water pollution can be caused by phosphates that are added to detergents. The phosphorus in a 1.60 g sample of washing powder was precipitated as $Mg_2P_2O_7$. The precipitate was filtered and dried and had a mass of 0.0632 g.

a. Calculate the:

i. amount, in mol, of the $Mg_2P_2O_7$ precipitate 1 mark

ii. mass, in grams, of phosphorus in the sample 2 marks

iii. percentage of phosphorus in the sample 1 mark

b. Explain what the effect would be on the result if the precipitate was not completely dried. 2 marks

ISBN 978 0 6557 0015 9

EXAM-STYLE QUESTIONS

Question 5

The concentration of iron in bore water was determined by oxidising the iron to Fe^{3+} and then complexing the Fe^{3+} ions with SCN^- ions, to form red-coloured $FeSCN^{2+}$ ions.

A sample of the bore water was then placed in a colorimeter equipped with a blue-green light filter and its absorbance was measured. The absorbances of three standard solutions were also measured, as shown in the table.

Iron standards ($mg\,mL^{-1}$)	Absorbance
2	3.0
4	5.8
6	8.8
sample	4.4

a. Construct a calibration curve of absorbance versus concentration of iron. 3 marks

b. Determine the concentration of iron in the sample in $mg\,mL^{-1}$. 1 mark

c. Explain why a blue-green filter is used, rather than a red filter. 2 marks

UNIT 2

How do chemical reactions shape the natural world?

AREA OF STUDY 3

How do quantitative scientific investigations develop our understanding of chemical reactions?

Outcome

On completion of this unit the student should be able to draw an evidence-based conclusion from primary data generated from a student-adapted or student-designed scientific investigation related to the production of gases, acid–base or redox reactions or the analysis of substances in water.

Key knowledge

Investigation design

- chemical science concepts specific to the selected scientific investigation and their significance, including the definition of key terms
- scientific methodology relevant to the selected scientific investigation, selected from the following: classification and identification; controlled experiment; fieldwork; modelling; product, process or system development; or simulation
- techniques of primary qualitative and quantitative data generation relevant to the investigation
- accuracy, precision, repeatability, reproducibility, resolution, and validity of measurements in relation to the investigation
- health, safety and ethical guidelines relevant to the selected scientific investigation

Scientific evidence

- the distinction between an aim, a hypothesis, a model, a theory and a law
- observations and investigations that are consistent with, or challenge, current scientific models or theories
- the characteristics of primary data
- ways of organising, analysing and evaluating generated primary data to identify patterns and relationships, and to identify sources of error
- the use of a logbook to authenticate generated primary data
- the limitations of investigation methodologies and methods, and of data generation and/or analysis

Science communication

- the conventions of scientific report writing, including scientific terminology and representations, standard abbreviations and units of measurement
- ways of presenting key findings and implications of the selected scientific investigation

Investigating the concentration of ethanoic acid in different types of vinegar

You will adapt or design and then conduct a scientific investigation to generate appropriate quantitative data, organise and interpret the data, and reach a conclusion in response to the research question.

The investigation will relate to the production of gases, acid–base or redox reactions, or the analysis of substances in water, and draws on key knowledge and related key science skills from Areas of Study 1 and/or 2.

ASSESSMENT FOR OUTCOME 3

A report of a student-adapted or student-designed scientific investigation using a selected format such as a scientific poster, an article for a scientific publication, a practical report, an oral presentation, a multimedia presentation or a visual representation.

SUGGESTED DURATION

A minimum of 7 hours of class time should be devoted to undertaking and communicating findings related to Area of Study 3. As per Areas of Study 1 and 2, your teacher will be your primary guide. To assist you, the key steps to follow, relevant scientific investigation sections and suggested time allocation are set out below.

The sample investigation on the following page steps you through a controlled experiment, conducted following the designing and planning phase of the investigation. The suggested duration for this part of the investigation includes 15–20 minutes for set-up, and 90 minutes for observation and recording.

USING THIS GUIDE

There is scope for developing an investigation on a range of topics relevant to Unit 2. This sample investigation is drawn from the key knowledge related to analysis for acids and bases as addressed in Unit 2 Area of Study 2. In particular, it further explores the concentration of ethanoic acid in different types of vinegar.

Refer to the Toolkit at the beginning of this book for more detailed information on designing and conducting scientific investigations, and presenting a scientific report.

Suggested time allocations for your investigation			
Scientific investigation section	**Key step**	**Relevant poster section(s)**	**Suggested time allocation (min)**
Designing and planning your investigation	Step 1: Developing aims, hypotheses and predictions	Title Introduction	60–120
	Step 2: Determining appropriate methodology and methods	Methodology and methods	60–120
Conducting your investigation and recording and presenting data	Step 3: Conducting your investigation to generate primary data		120–240
	Step 4: Recording, organising and presenting your data	Results	120–180
Discussing your investigation and drawing evidence-based conclusions	Step 5: Analysing and evaluating your data	Discussion Conclusion	60–120
	Step 6: Referencing	References and acknowledgements	30
Reporting on your investigation	Step 7: Preparing your scientific report	All sections	60–180

 ISBN 978 0 6557 0015 9

SAMPLE INTRODUCTION

BACKGROUND

Vinegar is a household substance used for cooking, salad dressings and cleaning. The active ingredient in vinegar is ethanoic acid, CH_3COOH, a weak monoprotic acid, which gives vinegar its sharp, acidic taste.

There are a number of different types of vinegar available for different taste preferences and uses. These include:

- white vinegar
- balsamic vinegar
- white wine vinegar
- red wine vinegar
- apple cider vinegar
- rice wine vinegar.

The question under investigation is: *Do different types of vinegar contain different concentrations of ethanoic acid?*

Volumetric analysis will be carried out using the strong base sodium hydroxide. Sodium hydroxide and the ethanoic acid in vinegar react according to the equation:

$$CH_3COOH(aq) + NaOH(aq) \rightarrow CH_3COONa(aq) + H_2O(l)$$

AIM

To investigate the concentration of ethanoic acid in different types of vinegar

HYPOTHESIS

Different types of vinegar will have different concentrations of ethanoic acid because the different sources of ethanoic acid have different levels of alcohol.

METHODOLOGY AND METHODS

Remember, your methodology is broader than the methods you have selected for this investigation. It includes a rationale for the approach taken and why this is important to the investigation. The methods are the specific steps taken to generate data during your investigation. You will also need to list all of the materials required to conduct the investigation. The methodology for the sample investigation is a controlled experiment.

METHODS

Several factors need to be considered when planning the method for an investigation. These may include:

- the independent and dependent variables
- how to control all other variables in the investigation
- how the brand of vinegar may also affect ethanoic acid concentration.

Prepare step-by-step instructions. Remember that these instructions should be clear and easy for someone else to follow.

RESULTS

The results of the investigation will include quantitative data collected, along with other relevant information. This may include:

- photographs and a description of the vinegar samples used in the investigation, noting the type and brand
- photographs and a description of the experimental setup used in the investigation
- notes of observations recorded during the investigation.
- notes of errors and uncertainties relevant to the investigation.

It is critical to use a logbook to authenticate generated primary data and observations, and record any limitations that become apparent. Use pen and cross out errors; do not use any form of correction fluid or pull out pages. This is to demonstrate the full scientific process of your investigation throughout the experiment.

Remember to present your data/evidence in appropriate formats to illustrate trends, patterns and/or relationships. Think carefully about how best to present your findings.

DISCUSSION

The discussion is dedicated to interpretation and evaluation of primary data collected during the investigation. The discussion is also where you will evaluate the methodology and methods used and possible sources of experimental errors/mistakes or bias. You should address accuracy, precision, repeatability, resolution and validity of measurements. A discussion of limitations should consider how identified factors have impacted on the investigation, as well as suggestions about how to reduce or eliminate these in subsequent investigations.

Results should be cross-referenced to relevant chemical concepts. Results should also be linked to the investigation question and aim, and explain whether or not the data supports your hypothesis.

The investigation may prompt suggestions for further investigation, and these should also be outlined in the discussion.

CONCLUSION

The conclusion is a clear and concise statement that summarises the findings of the investigation and responds to the investigation question. The conclusion should state whether the hypothesis is supported or refuted, drawing on evidence from the investigation to support this. No new information should be introduced.

REFERENCES AND ACKNOWLEDGEMENTS

Remember to use in-text citations as appropriate, and reference and acknowledge all secondary sources used in the course of planning and conducting your investigation. This may include:

- your logbook—raw data and calculations
- worksheets supplied by your teacher
- school website—support material
- field guides
- internet sources.

When referencing published sources, it is important to provide bibliographical data according to accepted protocols, such as the APA style. For example: Chan, D., Commons, C., Commons, P., Derry, L., Freer, E., Huddart, E., Lennard, L., MacEoin, M., Moylan, M., O'Shea, P., Ross, R., & Vanderkruk, K. (2023). *Heinemann Chemistry 1* (6th ed.). Pearson Australia.

Periodic table

Groups

Periods

s BLOCK — p BLOCK — d BLOCK

Key:

ATOMIC NUMBER → 12	24.31 ← RELATIVE ATOMIC MASS () indicates most stable isotope
SYMBOL → Mg	1090 ← BOILING POINT °C 650 ← MELTING POINT °C 1.3 ← ELECTRONEGATIVITY
ELECTRON STRUCTURE → $[Ne]3s^2$ Magnesium	

1 1.008
H -252.9 -259 2.2
$1s^1$ Hydrogen

Period	1	2	3	4	5	6	7	8	9	10	11	12	13	14	15	16	17	18
1																		2 4.003 He -268.9 -272 – $1s^2$ Helium
2	3 6.941 Li 1342 180 1.0 $1s^22s^1$ Lithium	4 9.012 Be 2468 1287 1.6 $1s^22s^2$ Beryllium											5 10.81 B 4000 2077 2.0 $1s^22s^22p^1$ Boron	6 12.01 C 4827 3500 2.6 $1s^22s^22p^2$ Carbon	7 14.01 N -195.8 -210 3.0 $1s^22s^22p^3$ Nitrogen	8 16.00 O -183 -219 3.4 $1s^22s^22p^4$ Oxygen	9 19.00 F -188.1 -220 4.0 $1s^22s^22p^5$ Fluorine	10 20.18 Ne -246 -249 – $1s^22s^22p^6$ Neon
3	11 22.99 Na 882.9 97.8 0.9 $[Ne]3s^1$ Sodium	12 24.31 Mg 1090 650 1.3 $[Ne]3s^2$ Magnesium											13 26.98 Al 2519 660.3 1.6 $[Ne]3s^23p^1$ Aluminium	14 28.09 Si 3265 1414 1.9 $[Ne]3s^23p^2$ Silicon	15 30.97 P 280.5 44.2 2.2 $[Ne]3s^23p^3$ Phosphorus	16 32.07 S 444.6 115.2 2.6 $[Ne]3s^23p^4$ Sulfur	17 35.45 Cl -34 -102 3.2 $[Ne]3s^23p^5$ Chlorine	18 39.95 Ar -185.8 -189 – $[Ne]3s^23p^6$ Argon
4	19 39.10 K 758.8 63.4 0.8 $[Ar]4s^1$ Potassium	20 40.08 Ca 1484 842 1.0 $[Ar]4s^2$ Calcium	21 44.96 Sc 2836 1541 1.4 $[Ar]3d^14s^2$ Scandium	22 47.87 Ti 3287 1670 1.5 $[Ar]3d^24s^2$ Titanium	23 50.94 V 3407 1910 1.6 $[Ar]3d^34s^2$ Vanadium	24 52.00 Cr 2671 1907 1.7 $[Ar]3d^54s^1$ Chromium	25 54.94 Mn 2061 1246 1.6 $[Ar]3d^54s^2$ Manganese	26 55.85 Fe 2861 1538 1.8 $[Ar]3d^64s^2$ Iron	27 58.93 Co 2927 1495 1.9 $[Ar]3d^74s^2$ Cobalt	28 58.69 Ni 2913 1455 1.9 $[Ar]3d^84s^2$ Nickel	29 63.55 Cu 2560 1085 1.9 $[Ar]3d^{10}4s^1$ Copper	30 65.38 Zn 907 419.5 1.6 $[Ar]3d^{10}4s^2$ Zinc	31 69.72 Ga 2229 27.8 1.8 $[Ar]3d^{10}4s^24p^1$ Gallium	32 72.64 Ge 2833 938.2 2.0 $[Ar]3d^{10}4s^24p^2$ Germanium	33 74.92 As 613 816.8 2.2 $[Ar]3d^{10}4s^24p^3$ Arsenic	34 78.96 Se 684.8 220.8 2.6 $[Ar]3d^{10}4s^24p^4$ Selenium	35 79.90 Br 58.8 -7.1 3.0 $[Ar]3d^{10}4s^24p^5$ Bromine	36 83.80 Kr -153.4 -157 – $[Ar]3d^{10}4s^24p^6$ Krypton
5	37 85.47 Rb 687.8 39 0.8 $[Kr]5s^1$ Rubidium	38 87.61 Sr 1377 769 1.0 $[Kr]5s^2$ Strontium	39 88.91 Y 3345 1522 1.2 $[Kr]4d^15s^2$ Yttrium	40 91.22 Zr 4406 1854 1.3 $[Kr]4d^25s^2$ Zirconium	41 92.91 Nb 4741 2477 1.6 $[Kr]4d^45s^1$ Niobium	42 95.96 Mo 4639 2622 2.2 $[Kr]4d^55s^1$ Molybdenum	43 (98) Tc 4262 2157 2.1 $[Kr]4d^55s^2$ Technetium	44 101.1 Ru 4147 2333 2.2 $[Kr]4d^75s^1$ Ruthenium	45 102.9 Rh 3695 1963 2.3 $[Kr]4d^85s^1$ Rhodium	46 106.4 Pd 2963 1555 2.2 $[Kr]4d^{10}5s^0$ Palladium	47 107.9 Ag 2162 961.8 1.9 $[Kr]4d^{10}5s^1$ Silver	48 112.4 Cd 766.8 321.1 1.7 $[Kr]4d^{10}5s^2$ Cadmium	49 114.8 In 2072 156.6 1.8 $[Kr]4d^{10}5s^25p^1$ Indium	50 118.7 Sn 2602 231.9 2.0 $[Kr]4d^{10}5s^25p^2$ Tin	51 121.8 Sb 1587 630.6 2.0 $[Kr]4d^{10}5s^25p^3$ Antimony	52 127.6 Te 987.8 449.5 2.1 $[Kr]4d^{10}5s^25p^4$ Tellurium	53 126.9 I 184.4 113.7 2.7 $[Kr]4d^{10}5s^25p^5$ Iodine	54 131.3 Xe -108.1 -112 2.6 $[Kr]4d^{10}5s^25p^6$ Xenon
6	55 132.9 Cs 670.8 28.5 0.8 $[Xe]6s^1$ Caesium	56 137.3 Ba 1845 727 0.9 $[Xe]6s^2$ Barium	57 138.9 La 2464 920 1.1 $[Xe]5d^16s^2$ Lanthanum	72 178.5 Hf 4600 2233 1.3 $[Xe]4f^{14}5d^26s^2$ Hafnium	73 180.9 Ta 5455 3017 1.5 $[Xe]4f^{14}5d^36s^2$ Tantalum	74 183.9 W 5555 3414 1.7 $[Xe]4f^{14}5d^46s^2$ Tungsten	75 186.2 Re 5596 3454 1.9 $[Xe]4f^{14}5d^56s^2$ Rhenium	76 190.2 Os 5008 3033 2.2 $[Xe]4f^{14}5d^66s^2$ Osmium	77 192.2 Ir 4428 2446 2.2 $[Xe]4f^{14}5d^76s^2$ Iridium	78 195.1 Pt 3825 1768 2.2 $[Xe]4f^{14}5d^96s^1$ Platinum	79 197.0 Au 2836 1064 2.4 $[Xe]4f^{14}5d^{10}6s^1$ Gold	80 200.6 Hg 356.6 -38.8 1.9 $[Xe]4f^{14}5d^{10}6s^2$ Mercury	81 204.4 Tl 1473 303.8 1.8 $[Xe]4f^{14}5d^{10}6s^26p^1$ Thallium	82 207.2 Pb 1749 327.5 1.8 $[Xe]4f^{14}5d^{10}6s^26p^2$ Lead	83 209.0 Bi 1564 271.4 1.9 $[Xe]4f^{14}5d^{10}6s^26p^3$ Bismuth	84 (210) Po 962 253.8 2.0 $[Xe]4f^{14}5d^{10}6s^26p^4$ Polonium	85 (210) At 366.8 301.8 2.2 $[Xe]4f^{14}5d^{10}6s^26p^5$ Astatine	86 (222) Rn -61.8 -71.2 – $[Xe]4f^{14}5d^{10}6s^26p^6$ Radon
7	87 (223) Fr 676.8 27 0.7 $[Rn]7s^1$ Francium	88 (226) Ra 1140 699.8 0.9 $[Rn]7s^2$ Radium	89 (227) Ac 3200 1050 1.1 $[Rn]6d^17s^2$ Actinium	104 (261) Rf – – – $[Rn]5f^{14}6d^27s^2$ Rutherfordium	105 (262) Db – – – $[Rn]5f^{14}6d^37s^2$ Dubnium	106 (266) Sg – – – $[Rn]5f^{14}6d^47s^2$ Seaborgium	107 (264) Bh – – – $[Rn]5f^{14}6d^57s^2$ Bohrium	108 (267) Hs – – – $[Rn]5f^{14}6d^67s^2$ Hassium	109 (268) Mt – – – $[Rn]5f^{14}6d^77s^2$ Meitnerium	110 (271) Ds – – – $[Rn]5f^{14}6d^87s^2$ Darmstadtium	111 (272) Rg – – – $[Rn]5f^{14}6d^97s^2$ Roentgenium	112 (285) Cn – – – $[Rn]5f^{14}6d^{10}7s^2$ Copernicium	113 (284) Nh – – – $[Rn]5f^{14}6d^{10}7s^27p^1$ Nihonium	114 (289) Fl – – – $[Rn]5f^{14}6d^{10}7s^27p^2$ Flerovium	115 (289) Mc – – – $[Rn]5f^{14}6d^{10}7s^27p^3$ Moscovium	116 (292) Lv – – – $[Rn]5f^{14}6d^{10}7s^27p^4$ Livermorium	117 (294) Ts – – – $[Rn]5f^{14}6d^{10}7s^27p^5$ Tennessine	118 (294) Og – – – $[Rn]5f^{14}6d^{10}7s^27p^6$ Oganesson

f BLOCK

58 140.1 Ce 3443 799 1.1 $[Xe]4f^25d^06s^2$ Cerium	59 140.9 Pr 3520 931 1.1 $[Xe]4f^35d^06s^2$ Praseodymium	60 144.2 Nd 3074 1016 1.1 $[Xe]4f^45d^06s^2$ Neodymium	61 (145) Pm 3000 1042 – $[Xe]4f^55d^06s^2$ Promethium	62 150.4 Sm 1794 1072 1.1 $[Xe]4f^65d^06s^2$ Samarium	63 152.0 Eu 1794 1072 – $[Xe]4f^75d^06s^2$ Europium	64 157.3 Gd 3273 1313 1.2 $[Xe]4f^75d^16s^2$ Gadolinium	65 158.9 Tb 3230 1359 – $[Xe]4f^95d^06s^2$ Terbium	66 162.5 Dy 2567 1412 1.2 $[Xe]4f^{10}5d^06s^2$ Dysprosium	67 164.9 Ho 2700 1472 1.2 $[Xe]4f^{11}5d^06s^2$ Holmium	68 167.3 Er 2868 1529 1.2 $[Xe]4f^{12}5d^06s^2$ Erbium	69 168.9 Tm 1950 1545 1.3 $[Xe]4f^{13}5d^06s^2$ Thulium	70 173.1 Yb 1196 824 – $[Xe]4f^{14}5d^06s^2$ Ytterbium	71 175.0 Lu 3402 1663 1.0 $[Xe]4f^{14}5d^16s^2$ Lutetium
90 232.0 Th 4875 1750 1.3 $[Rn]5f^06d^27s^2$ Thorium	91 231.0 Pa – 1572 1.5 $[Rn]5f^26d^17s^2$ Protactinium	92 238.0 U 4131 1135 1.7 $[Rn]5f^36d^17s^2$ Uranium	93 (237) Np 3902 644 1.3 $[Rn]5f^46d^17s^2$ Neptunium	94 (244) Pu 3228 640 1.3 $[Rn]5f^66d^07s^2$ Plutonium	95 (243) Am 2011 1176 – $[Rn]5f^76d^07s^2$ Americium	96 (247) Cm 3110 1340 – $[Rn]5f^76d^17s^2$ Curium	97 (247) Bk – 986 – $[Rn]5f^96d^07s^2$ Berkelium	98 (251) Cf – 900 – $[Rn]5f^{10}6d^07s^2$ Californium	99 (252) Es – – – $[Rn]5f^{11}6d^07s^2$ Einsteinium	100 (257) Fm – – – $[Rn]5f^{12}6d^07s^2$ Fermium	101 (258) Md – 827 – $[Rn]5f^{13}6d^07s^2$ Mendelevium	102 (259) No – – – $[Rn]5f^{14}6d^07s^2$ Nobelium	103 (262) Lr – – – $[Rn]5f^{14}6d^17s^2$ Lawrencium